A NATURALIST'S

BIRDS
OF
HONG KONG

Ray Tipper

JOHN BEAUFOY PUBLISHING

Reprinted with corrections 2018
First published in the United Kingdom in 2016 by John Beaufoy Publishing Ltd
11 Blenheim Court, 316 Woodstock Road, Oxford OX2 7NS, England
www.johnbeaufoy.com

10 9 8 7 6 5 4 3 2

Copyright © 2016, 2018 John Beaufoy Publishing Limited
Copyright in text © 2016, 2018 Ray Tipper
Copyright in photographs © 2016, 2018 see below

Copyright in maps © 2016, 2018 John Beaufoy Publishing Limited

Photo Credits
Front cover: *main* Eurasian Hoopoe, *bottom left* Daurian Redstart, *bottom centre* Red-necked Phalarope, *bottom right* Eastern Yellow Wagtail © *Ray Tipper*. **Back cover:** Bonelli's Eagle © Ray Tipper.
Title page: Japanese White-eye © Ray Tipper.
Contents page: Curlew Sandpiper © Ray Tipper.
Main descriptions: All photos by **Ray Tipper** except as detailed below. Photos are denoted by a page number followed by t (top), b (bottom), l (left), r (right) or c (centre).
Tom Beeke 88tl, 88b; **Abdelhamid Bizid** 102b, 110b, 118b; **Allen Chan** 78b; **Amar-Singh HSS** 76b, 77t, 83t, 113t, 130t; **Bonnie Chan** 35t, 65b, 66t, 70tl, 81t, 106bl, 106br, 123t, 123b, 126t, 130b, 132t, 139b; **Sam Chan** 110t; **Louise Cheung** 91t, 136t, 141b; **Chung Wing-kin** 49t, 61t, 99t, 120b, 134b, 155b; **Chung Yun Tak** 69br, 148b; **Kin Yip Ho** 16b, 20b, 24t, 33t, 45bl, 97b, 107b, 136bl, 136br, 138t, 157tl, 157tr; **John & Jemi Holmes** 21b, 80t, 80b, 81b, 82t, 92tl, 124b, 125t, 135t; **Herman Ip** 109t; **Ajuwat Jearwattanakanok** 94t, 129t; 阿棋 **Looking@ Nature** 86b, 96b, 97t, 101b, 105b, 106t, 108b, 125b, 127t, 138b, 140tl, 155t; **Koel Ko** 13cl, 51b, 75b, 85b, 117t, 137tr; **Henry Lui** 25t, 77bl, 77br, 79b, 90t, 92tr, 121t, 121b, 122t, 124t, 140b, 145b, 156t; **Du Qing** 13tr, 13br; **Ivan Tse** 21t, 32b, 36t, 37t, 69tl, 69tr, 90b, 108t, 114t, 114b, 144t; **Martin Williams** 34b; **Michelle and Peter Wong** 12tl, 13tl, 18b, 35b, 39b, 47b, 57b, 65t, 79t, 84t, 89t, 91b, 92bl, 92br, 96t, 103tl, 109b, 113b, 115t, 142t, 146tl, 146br, 147b, 157b; **Yong Ding Li** 133t; **Yu Yat-tung** 72b, 154t.

All rights reserved. No part of this publication may be reproduced, stored in a retrieval system or transmitted in any form or by any means, electronic, mechanical, photocopying, recording or otherwise, without the prior written permission of the Publisher.

The presentation of material in this publication and the geographical designations employed do not imply the expression of any opinion whatsoever on the part of the Publisher concerning the legal status of any country, territory or area, or concerning the delimitation of its frontiers or boundaries.

ISBN 978-1-909612-08-2

Edited by Krystyna Mayer

Designed by Gulmohur Press, New Delhi

Printed and bound in Malaysia by Times Offset (M) Sdn. Bhd.

·Contents·

Acknowledgements 4

Scope of book 4

Taxonomy and nomenclature 4

Climate 4

Habitat 5

Where to go 5

Waders 11

Chinese Endemics in Hong Kong 12

Submission of records 14

Glossary 14

Bird topography 15

Species Accounts and Photographs 16

Checklist of the Birds of Hong Kong 158

Further Information 172

Index 174

Introduction

ACKNOWLEDGEMENTS

I would like to express my sincere thanks to Mathew Cheng, Reserve Manager of Mai Po Nature Reserve, who made arrangements for my frequent visits to the reserve over a seven-week period at the beginning of 2014. John Allcock and Bena Smith kept me updated on the location of species I wanted to photograph. Paul Leader accompanied me on several days in the search for targeted species to photograph. My special thanks go to Geoff Welch, who assisted in all manner of ways, including advising on species abundance and status in Hong Kong as well as reviewing and commenting on the first draft of this book, a task that was also undertaken to great effect by Mike Chalmers. His knowledge and comments were central to improving the species accounts. Yu Yat Tung was immensely helpful and invested an enormous amount of time in successfully sourcing many of the photographs I was unable to provide from my personal portfolio. He also reviewed and corrected the Chinese names. Gary Chow kindly advised and supplied Chinese names of species recently new to Hong Kong. Bonnie Chan, Sam Chan, Louis Cheun, Johnson Chung, Chung Wing King, Chung Yun Tak, Kinni Ho, Jemi and John Holmes, Herman Ip, Ivan Tse, Koel Ko, Lau Sin Pang, Martin Williams, Peter and Michelle Wong, Yu Yat Tung, and Yong Ding Li donated photographs (see p.2). Finally, I am indebted to Richard Bryson, who generously provided me with accommodation for the whole of my visit to Hong Kong in 2014.

SCOPE OF THIS BOOK

As at the beginning of September 2017, the bird list of the Special Administrative Region of Hong Kong stood at 549 species, a rise of 14 in a little over a year since the first publication of this book. Close to half this total is represented by species that are vagrants, rare or, at best, scarce and unlikely to be seen during the course of a regular day's birdwatching. It is the remaining 282 species, whose status is described as Abundant, Common or Uncommon (and sometimes qualified by words such as 'fairly' or 'locally') in recent editions of the Hong Kong Bird Report, that form the subject of this guide.

TAXONOMY AND NOMENCLATURE

The taxonomy, species sequence and English, Chinese and scientific names follow those adopted by the Hong Kong Bird Watching Society as published on its website (see 'Where to go', below). In one instance, for the Red-throated Flycatcher a widely adopted alternative name, Taiga Flycatcher, is preferred.

CLIMATE

Hong Kong enjoys distinct seasons linked to the regular movements of the East Asian Monsoon system. January and February, typically with northerly air flows, are the coldest months, when temperatures can fall below 10 °C, with frost occasionally forming on high ground. In March the influence of northern weather systems lessens, and warmer air gives rise to mist, light rain and an increase in humidity typical of the onset of spring. Winds turn to easterlies. The period from May to September is characterized by summer

■ Habitat ■

rain and hot, oppressive conditions. Temperatures commonly exceed 30 °C and the relative humidity, broadly speaking, stands at 80 per cent or higher. Hardly ever does a summer go by without Hong Kong having to close down for the passage of a typhoon or two. September sees the return of the north-east monsoon, and the final quarter of the year experiences settled conditions with a southerly airflow producing dry and sunny weather. (With acknowledgement to Lam, C. Y., in Carey et al., 2001, *The Avifauna of Hong Kong*.)

Habitat

Hong Kong represents a tiny region of the People's Republic of China. With a total land area of a trifle over 1,100sq km spread over more than 250 islands, and a coastline measuring some 730km, Hong Kong abuts mainland China in the north but is otherwise surrounded by the South China Sea. It is hilly in character yet nowhere does the elevation reach 1,000 m. Tai Mo Shan, close to the centre of the New Territories, rises to 957m, the highest point in Hong Kong. Originally, Hong Kong was clothed in broadleaved woodland, but it is believed that this had largely disappeared at the hands of humans by the middle of the 18th century. Today, human influence on the environment is evident everywhere, which is hardly surprising given that the population exceeds 7.4 million and continues to grow. Residential and industrial buildings, together with associated structures such as roads and reservoirs, account for more than 27 per cent of the total land area. Grass, shrub and woodland occupy more than 66 per cent, an impressively large percentage, though most is rough hillside that holds limited attraction to birdlife. Some 6 per cent of the land is utilized for agriculture and pisciculture, both significantly important avian habitats.

From a birding perspective, habitat in Hong Kong can be described broadly as wetlands, woodland, open lowland, upland grassland, and artificial and urban areas. Wetlands, unarguably, provide the largest concentrations of birds, both numerically and in terms of species. The coastline in the north has impressive intertidal mudflats, while many of the islands have rocky shorelines. Commercial fish ponds and, to a lesser extent, *gei wai* (traditional shrimp farms that have all but disappeared) are an important feature of the north-west, and open water occurs in the form of reservoirs. Woodland has regenerated since the end of the Second World War, and comparatively large areas of secondary broadleaved forest are gradually maturing. Stands of mangrove, especially around Deep Bay, form a cross-over between wetland and woodland. Open lowland habitat features farmland, both worked and abandoned, shrubbery and grassland. On the hilltops grassland prevails, with strips of tree growth following the incisions of watercourses. The general lack of trees on these grasslands is a result of recurring dry-weather fires that inhibit their growth. Urban areas, in particular the many small to sizable wooded parks, attract a surprising number of species, some of which, such House Swift, Barn Swallow, Red-whiskered and Chinese Bulbuls, Crested Myna, Black-collared Starling and Eurasian Tree Sparrow, are just as likely to be seen in built-up areas as anywhere else in Hong Kong.

Where to go

The website of the Hong Kong Bird Watching Society, www.hkbws.org.hk, features a comprehensive compilation of both major and lesser birdwatching sites in Hong Kong in detail, with instructions on how to reach them by public transport. Here are listed perhaps the five most rewarding.

MAI PO NATURE RESERVE

This is without doubt the premier birdwatching site in Hong Kong. The mix of intertidal mudflats, mangrove woodland, fish ponds, reed beds, an artificial 'scrape' and tree-lined paths makes for an unrivalled range of habitats that attract an exceptional cross-section of bird species. Designated a Ramsar site in 1999, Mai Po is of international significance and is especially important as a refuge for long-distance migrant waders on the East Asian–Australasian Shorebird Flyway, which stretches from New Zealand to the Russian Far East. A series of hides includes a tower and several floating hides at the ends of boardwalks winding through the mangroves that overlook the mudflats of Deep Bay. Winter brings massive influxes of ducks, waders and gulls, and a significant proportion of the world's Black-faced Spoonbills, a globally endangered species. In both spring and autumn, in addition to the migrant wader spectacle, migrant passerines are evident. Some, like Siberian Rubythroat, Red-flanked Bluetail, Asian Brown Flycatcher and Eastern Yellow Wagtail, remain throughout the winter.

Fish ponds and Great Cormorant roost Mai Po NR.

WHERE TO GO

Waterbirds, principally Pied Avocets, in Deep Bay, Mai Po NR, with Shenzhen in the background.

Visits to the reserve, which is managed by WWF Hong Kong, are strictly controlled, and entry is only possible to holders of the necessary visitor permits. Current requirements for the issue of permits are detailed on the Hong Kong Bird Watching Society's website. The state of the tide is of considerable importance when choosing a time to visit the boardwalk hides. At low tide birds are very distant and at high tide few birds are present. It is therefore important to coincide one's visit with a rising tide, ideally of at least 2m at its high point.

'LONG VALLEY' AND HO SHEUNG HEUNG

Located in the north-west New Territories within walking distance of Sheung Shui MTR station, the Long Valley site occupies a unique area of agricultural land, intensively worked as both wet and dry vegetable fields. A few abandoned fish ponds and marshy areas add variety to the habitat, which is further enhanced by the adjacent *fung shui* woods of Ho Sheung Heung. Thrushes, chats, pipits and wagtails (including occasional Buff-bellied Pipits and Citrine Wagtails) are all winter regulars at Long Valley. It is one of the most reliable sites for seeing Greater Painted-snipe and, with luck, one of the less common cuckoos, while Ho Sheung Heung attracts warblers, flycatchers and flowerpeckers. Paths make access and coverage of the whole area straightforward.

■ Where to go ■

A rare sight in Hong Kong – the agricultural fields of Long Valley.

TAI PO KAU NATURE RESERVE

Lying to the south of the town of Tai Po in the central New Territories, Tai Po Kau is the most extensive stand of woodland in Hong Kong. Designated a nature reserve and managed by the Agriculture, Fisheries and Conservation Department of the Hong Kong Government, a series of well-maintained and signposted walks gives excellent access to the forest. Throughout the year this is the best site in Hong Kong for woodland birds, although it requires a good deal of patience to see many of the species present. Among the more easily located is the striking Velvet-fronted Nuthatch, just one of a clutch of non-native species that have become established here.

TAI MO SHAN

The highest point in Hong Kong tops a largely grass-covered hillside with metalled road access. It is a reliable site for Chinese Francolin (easily heard but seldom seen), Chinese Grassbird and Upland Pipit.

PO TOI ISLAND

Only in recent years has Po Toi's full potential as a migration hotspot been fully recognized. Located an hour's ferry journey south of Hong Kong Island, Po Toi is well positioned to receive tired migrants and regularly produces 'firsts for Hong Kong'. The ferry ride offers opportunities to see seabirds at the appropriate times of year.

Where to go

Tai Mo Shan

Tai Po Kau

■ Where to go ■

Po Toi Island

At the end of the Habitat, Habits and Status section of each species account, the following alphabetical code has been adopted to provide a general guide as to where that species might most likely be seen.

M   Mai Po Nature Reserve
L   Long Valley
T   Tai Po Kau
H   Parks and gardens
I   Offshore islands

Waders

Above all else, Hong Kong is noted for its waders. No fewer than 60 species have been recorded, of which just five breed or have bred in the territory. Black-winged Stilt, Little Ringed Plover and Greater Painted-snipe regularly breed in small numbers. Pied Avocet appears to be in the process of establishing itself as a breeding species (with several breeding attempts in recent years) while Pheasant-tailed Jacana, which was last recorded breeding in 1974, could well return as habitat management in the north-west New Territories seeks to provide more suitable conditions for the species.

▪ WADERS ▪

Winter flocks on the intertidal mudflats in Deep Bay can reach a total of more than 25,000, close to 60 per cent of which are likely to be made up of an amazing gathering of close to 15,000 Pied Avocets. Eurasian Curlew, Common Greenshank and Dunlin all have peaks of around 1,000 (and Kentish Plover nearly as many). Dunlin winter high counts have exceeded 3,000, and for Marsh Sandpiper more than 2,000 is the norm. Midwinter numbers of Pacific Golden and Grey Plovers, Black-tailed Godwits, and Spotted and Common Redshanks, although smaller are nevertheless noteworthy.

Nordmann's Greenshank preening.

It is the spring migration, however, that brings the real excitement as hundreds of waders migrating on the East Asian–Australasian Flyway pause in Deep Bay and Mai Po to rest and 'refuel'. Then, it is the breathtaking array of species that enthrals, with many individuals in stunning breeding plumage, including such essentially eastern waders as Far Eastern Curlew, Grey-tailed Tattler, Great Knot, Red-necked and Long-toed Stints, and Sharp-tailed Sandpiper. Topping the list, however, come three species, Asian Dowitcher, Nordmann's Greenshank and Spoon-billed Sandpiper, which were virtually unknown to Western birders before the establishment of Mai Po as a nature reserve, an event that in the 1980s and '90s witnessed an annual pilgrimage to Hong Kong by wader enthusiasts from all over the world.

CHINA ENDEMICS IN HONG KONG

The People's Republic of China boasts 59 endemic bird species and another 15 that are breeding endemics. Hong Kong, unsurprisingly in view of its small size, has no endemics of its own but no fewer than 13 of the 74 China endemics and breeding endemics have been recorded in the Special Administrative Region. These are: Swinhoe's Minivet (Br), Yellow-bellied Tit, Chinese Penduline Tit (Br) Chinese Leaf Warbler (Br *), Goodson's Leaf Warbler, Sulphur-breasted Warbler (Br), Marten's Warbler (*), Alström's Warbler (Br *) Manchurian Reed Warbler (Br), Red-billed Starling (Br), Chinese Thrush (Br *), Hodgson's Redstart (Br *), Slaty Bunting (*)

Key: Br China breeding endemic
* Vagrant to Hong Kong

▪ Endemics ▪

All are passerines. Six are vagrants having been recorded in Hong Kong on fewer than ten occasions. Just three feature in the main text, namely, Chinese Penduline Tit and Goodson's Leaf Warbler, both being common winter visitors, and Red-billed Starling, an abundant winter visitor that is increasingly recorded in summer and has bred. The remaining four are scarce or rare visitors.

Swinhoe's Minivet

Yellow-bellied Tit

Chinese Penduline Tit

ENDEMICS

Goodson's Leaf Warbler

Sulphur-breasted Warbler

Manchurian Reed Warbler

Red-billed Starling

Chinese Thrush

Hodgson's Redstart

■ Glossary ■

Submission of records

The Hong Kong Bird Watching Society has collected records for more than 50 years and encourages the submission of records of both common and rare species. A record sheet is available from the society's website, as too is a separate form designed for the submission of rarity records. All records can be forwarded to the society by email (hkbws@hkbws.org.hk). A comprehensive report on the birds of Hong Kong (the Hong Kong Bird Report) is published annually.

Glossary

Many plumage and structural features are defined in the bird topography diagram below and are not, therefore, included in the following:

axillaries The feathers covering the base of the underwing ('armpit').
congeners Members of the same genus.
brood parasite Bird species that habitually lays its eggs in the nest of another species, which then incubates the eggs and rears the young.
calidrid Member of genus *Calidris*.
carpal Joint at point or bend of folded wing ('the wrist').
corvid Species of crow family, or Corvidae.
distal Furthest point from body, for example tip of bill.
ear-coverts Group of feathers immediately behind and below eye.
first-winter plumage The plumage attained following a post-juvenile moult.
gape The opening formed at the base of the bill when the two mandibles are parted.
genus A category or grouping of related species.
gonys Ridge towards tip of lower mandible, most obvious in large gull species.
greater coverts Line of wing-coverts closest to flight feathers.
jizz Overall impression of shape and movement of a bird.
larid Species of gull, or Laridae.
lesser coverts Line of wing-coverts furthest from flight feathers.
mandibles The two (upper and lower) sections of the bill.
median coverts Middle line of wing-coverts.
mesial Line down centre of throat.
malar Stripe down side of throat from base of bill.
morph Alternative, distinct plumage within a species, which is genetically inherited.
moustachial Stripe from base of bill extending along lower borders of ear-coverts.
nuchal Relating to nape.
palmations Webs between three forwards-pointing toes.
primaries Outer flight feathers, ten in number in most species.
primary projection Length of primaries that extends beyond tips of tertials on folded wing.
proximal Part of an appendage (for example bill, wing, tail) closest to body.
rectrices Tail feathers.
remiges The main flight feathers; together, the primaries and secondaries.
rictal Pertaining to gape.

Bird Topography

secondaries The flight-feathers positioned between the primaries and the tertials varying from six to 40 in number depending on the species.
scapulars Group of feathers between mantle and wing.
spatulate Description of a bill with a flattened and rounded tip, as in spoonbills.
submoustachial Stripe between moustachial and malar.
supercilium Stripe across head immediately above eye.
tertials Innermost flight feathers.
wing-coverts The three tracts of wing feathers that overlie flight feathers

Bird Topography

■ Ducks ■

Gadwall ■ *Anas strepera* 赤膀鴨 46–56cm

DESCRIPTION Medium-sized, plain-coloured dabbling duck. Breeding male essentially dark grey. Only at close range is it apparent that head, mantle, flanks and especially breast are vermiculated black and white. Rear flanks and undertail-coverts black. Long scapulars grey, broadly edged with cinnamon-buff. White speculum often visible on birds at rest and conspicuous in flight below chestnut median coverts. Female undistinguished except for small white speculum. Brownish-grey head; neck, breast and upperparts blackish-brown edged with buff; belly white. Eclipse male and immature birds similar to female. Bill greyish-black in breeding male, and greyish with orange-yellow sides in female, immature and non-breeding male. **HABITAT, HABITS AND STATUS** Inhabits shallow wetlands. Uncommon winter visitor to fish ponds and marshes in Hong Kong, almost exclusively in Deep Bay area. (M)

Female

Male

Falcated Duck ■ *Anas falcata* 羅紋鴨 48–54cm

DESCRIPTION Large-headed, bulky-looking, medium-sized dabbling duck. Male has reddish-purple crown, iridescent velvety-green head, neck and luxuriant maned crest. Small white patch above bill and white throat, bordered below by single, narrow black and white bands. Breast, upperparts and underparts closely vermiculated black and white, producing varying shades of silvery-grey; breast heavily scalloped; elongated scapulars and long, drooping black tertials boldly edged with whitish. Undertail-coverts black with yellow-buff patch on side. Female and immature warm brown and black; head and neck greyer, lacking brown tones. **HABITAT, HABITS AND STATUS** Favours shallow waters. Uncommon but annual winter visitor to Deep Bay area. (M)

■ Ducks ■

Eurasian Wigeon ■ *Anas penelope* 赤頸鴨 45–51cm

DESCRIPTION Neat, medium-sized dabbling duck. Male has ruddy-chestnut head with yellowish-cream forehead and crown, vinous-tinged breast, densely vermiculated black and white (appear grey) upperparts and flanks, elongated, white-edged black scapulars, white underparts and black rear. In flight, shows conspicuous white patch on upperwing-coverts. Rather small bill blue-grey with black tip. Female and immature variably rufous- or grey-brown, with white underparts and wing-coverts. Voice: male's call a distinctive loud, high-pitched, whistling *whee-oo*. **HABITAT, HABITS AND STATUS** Coastal wetlands and marshes, especially tidal estuaries and bays. Gregarious. Abundant winter visitor to Deep Bay area. (M)

Female *Male*

Chinese Spot-billed Duck ■ *Anas zonorhyncha* 中華斑嘴鴨 58–63cm

DESCRIPTION Large, plain dabbling duck with striking face pattern. Body dull blackish-brown; breast paler brown, heavily spotted blackish. Pale sides to head and neck emphasized by dark brown crown, eye-stripe and short cheek-stripe from bill-base. Speculum blue. Bill black with bright orange-yellow tip and black nail. Legs orange. Sexes similar; female somewhat duller than male. **SIMILAR SPECIES** Burmese race of **Indian Spot-billed Duck** *A. poecilorhyncha haringtoni* is a rare visitor to Hong Kong and is distinguished by lack of cheek-bar, green speculum and white tertials, often visible at rest and conspicuous in flight. **HABITAT, HABITS AND STATUS** Wetlands with emergent vegetation. Uncommon; records in recent years largely confined to Deep Bay area. (M)

■ Ducks ■

Northern Shoveler ■ *Anas clypeata* 琵嘴鴨 44–52cm

DESCRIPTION Medium-sized, bulky dabbling duck with conspicuously large, spatulate bill. Male boldly patterned with bottle-green head and neck, black mantle and undertail-coverts, white breast and stern, and deep chestnut flanks and underparts. Female has typical mottled brown plumage of dabbling ducks, though is easily distinguished by bill shape unique among northern hemisphere species. **HABITAT, HABITS AND STATUS** Favours freshwater wetlands and tidal estuaries. Abundant winter visitor to Deep Bay area. (M)

Male    *Female*

Northern Pintail ■ *Anas acuta* 針尾鴨 51–66cm

DESCRIPTION Medium-sized dabbling duck with a rather slender outline. Drake has dark chocolate-brown head and neck, throat and rear neck. Underparts, breast and lower neck white, with white extending upwards towards nape as narrow stripe. Body vermiculated grey and white. Yellowish-cream patch in front of black undertail-coverts. Black and cream scapulars elongated and pointed. Black central-tail feathers extend as a point up to 10cm beyond body. Female mottled pale brown; distinguishable from other ducks by plainer head and neck, slender, more elegant, long-necked shape and longish blue-grey bill. **HABITAT, HABITS AND STATUS** Occupies shallow aquatic habitats. Gregarious. Abundant winter visitor to Deep Bay area. (M)

Male   *Female*

▪ Ducks ▪

Garganey ▪ *Anas querquedula* 白眉鴨 37–41cm

DESCRIPTION Small dabbling duck. Male strikingly patterned with bold white crescent across grey-brown head from in front of eye to nape. Buff neck and breast scalloped black. Mantle dark brown; elongated, drooping scapulars white, edged with black. Flanks vermiculated grey; underparts white with brown rear. Female mottled brown with subtly distinctive head pattern of round pale patch at base of bill, pale supercilium and darkish brown eye-stripe. Bill grey, rather long and somewhat flattened. **HABITAT, HABITS AND STATUS** Largely confined to shallow, freshwater wetlands with emergent vegetation. Normally in pairs or small groups, though in flocks of 200 or more on migration. Common passage migrant, especially in autumn, mainly in Deep Bay area. Also winters in small numbers. (M)

Eurasian Teal ▪ *Anas crecca* 綠翅鴨 34–38cm

DESCRIPTION Small dabbling duck. Male with cinnamon head; wide green stripe, narrowly bordered buff, from in front of eye sweeping across to nape. Upperparts and flanks finely vermiculated black and white, and appearing solid dark grey at a distance. Yellow-buff patch at rear with conspicuous black border. Breast buffish, finely spotted black. White horizontal line formed by scapulars can appear prominent or absent depending on feather positioning. Female mottled brown and black, lacking distinctive features. **HABITAT, HABITS AND STATUS** Wide range of aquatic environments, from lakes, ponds and marshes, to tidal bays and estuaries. In Hong Kong an abundant winter visitor to Deep Bay area. (M, L)

Female *Male*

▪ Ducks & Game Birds ▪

Tufted Duck ▪ *Aythya fuligula*
鳳頭潛鴨 40–47cm

DESCRIPTION Smallish, neat diving duck. Male black with white flanks and underparts. In favourable light head shows purple gloss. Black crest falling from nape diagnostic but not always apparent. Female dark brown; paler on flanks with distinctly shorter crest than male. Individual females show varying amount of white around bill-base, and some acquire white undertail-coverts. **HABITAT, HABITS AND STATUS** Favours freshwater lakes and reservoirs, diving regularly when feeding. Unlike dabbling ducks, runs on surface of water when taking flight. Sociable. Abundant winter visitor to Deep Bay area. (M)

Chinese Francolin ▪ *Francolinus pintadeanus* 中華鷓鴣 c. 33cm

DESCRIPTION Unmistakable, being the only francolin within range. Male has unique head pattern of black crown, eye-stripe and face-stripe encircling white cheeks. Broad orangey-brown supercilium reaches back to nape; chin and throat white. Upperparts and upper breast black, spotted white. Underparts white, heavily scalloped with black; vent and wing-coverts rufous-brown. Female a dull version of male; black plumage replaced by brown with copious black spots and barring. More often heard than seen. Voice: far-carrying, 5–6-syllable, loud, throaty *whi ah-ah ah-ah*. **HABITAT, HABITS AND STATUS** Grassy scrub particularly on hill slopes, where it is fairly common and widespread. Secretive except when male is advertising in spring by calling from prominent position. (I)

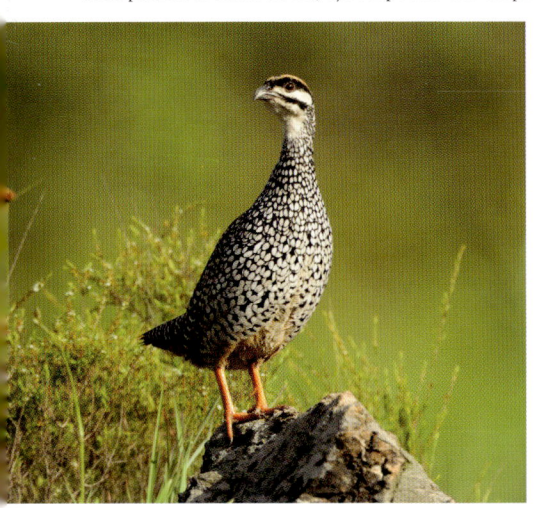

▪ Game Birds & Shearwater ▪

Japanese Quail ▪ *Coturnix japonica* 鵪鶉 17–19cm

DESCRIPTION Small, rotund game bird. Upperparts brown, cryptically streaked cream, cinnamon and black. Broad, creamy supercilium reaches neck. Underparts buff. Stubby blackish bill. Sexes are similar, with breeding male displaying more cinnamon tones on face and throat, and richer underparts than non-breeding male. Female and non-breeding male have double black bands at sides of throat and dark brown spotting on breast. **HABITAT, HABITS AND STATUS** Highly secretive ground dweller, rarely observed unless flushed. Uncommon autumn passage migrant, and occasional winter records in grassy and agricultural areas mainly in northern New Territories. (L)

Short-tailed Shearwater ▪ *Puffinus tenuirostris* 短尾鸌 40–43cm

DESCRIPTION Long, narrow, pointed winged, shortish tailed seabird. All-dark greyish-brown with variable whitish flash on underwing. Bill slim with tubular nostrils mounted on base of culmen. In flight, feet partially protrude beyond tail-tip. **SIMILAR SPECIES** Scarce **Streaked Shearwater** *Calonectris leucomelas* larger and longer-tailed, with paler, scaly brown upperparts and whitish head the extent of which variable. Underparts white, and underwings with brown border and streaking at carpal and across coverts. Vagrant **Wedge-tailed Shearwater** *P. pacificus* similar to Streaked but with all-dark underwings. Vagrant **Bulwer's Petrel** *Bulweria bulwerii* much the smallest, with notably long-pointed or wedge-shaped tail, and all-dark plumage except for pale upperwing-coverts bar. **HABITAT, HABITS AND STATUS** Pelagic seabird; flies close to water with rapid, stiff wingbeats followed by glide. Uncommon spring migrant recorded mainly in southern offshore waters. (I)

◾ GREBES ◾

Little Grebe ◾ *Tachybaptus ruficollis* 小鷿鷈 25–29cm

DESCRIPTION Smallish, plump waterbird with longish neck. Adult in breeding plumage has deep chestnut throat, cheeks and foreneck, black crown and hindneck, brown mantle and light brown flanks. Bill black with white tip and conspicuous yellow gape-flange. Iris yellow. Winter plumage lacks chestnut. Upperparts brown, foreneck and flanks light brown. Sexes are similar. Juvenile similar to winter adult, but with obvious broad pale stripes on head and hindneck. Voice: distinctive call a loud, whinnying trill. **HABITAT, HABITS AND STATUS** Strictly aquatic, preferring fish ponds and occasionally reservoirs to the sea. Makes lengthy dives for food; swims buoyantly. Common resident largely restricted to Deep Bay area. (M, L)

Great Crested Grebe ◾ *Podiceps cristatus* 鳳頭鷿鷈 46–51cm

DESCRIPTION Large grebe with long neck and sharply pointed, dagger-like bill. Winter plumage essentially black and white. Crown and flattened rear tufts black; hindneck and upperparts grey-brown; 'face', foreneck, breast and underparts white. Transformed in breeding plumage, which birds begin to acquire in Hong Kong from February, by presence of splendidly decorative burnt-orange merging to black plumes falling from sides of head to form a bushy frill. Crown tufts much enlarged. Bill grey-black fading to white tip (but pale pinkish in winter). Iris red. Sexes are similar; immature similar to winter adult. **HABITAT, HABITS AND STATUS** Exclusively aquatic, expert diver. Swims low in water with erect neck, which is drawn back onto body when relaxed. Invariably dives if disturbed in preference to flight. Common winter visitor to Deep Bay. (M)

SPOONBILLS

Eurasian Spoonbill ■ *Platalea leucorodia* 白琵鷺 80–90cm

DESCRIPTION Large white waterbird with long neck and legs; extraordinary long, spatulate black bill with deep yellow-orange patch of variable intensity at tip. In breeding plumage shaggy white crest, sometimes tinged yellowish-buff, hangs from nape, and yellow-buff half-collar forms at base of neck. Legs black. Sexes are similar. Immature as winter adult, but bill and legs pinkish, and tips of primaries black. **SIMILAR SPECIES** See Black-faced Spoonbill (below). **HABITAT, HABITS AND STATUS** Feeds on small fish and shrimps caught with a characteristic side-to-side sweeping motion of head and bill. Flies with neck outstretched. Gregarious, often in mixed flocks with Black-faced Spoonbill. Uncommon winter visitor to tidal mudflats, mangroves and fish ponds of Deep Bay. (M)

Black-faced Spoonbill ■ *Platalea minor* 黑臉琵鷺 60–78.5cm

DESCRIPTION Similar to and distinguished from much scarcer (in Hong Kong) Eurasian Spoonbill (see above) by smaller size, though this is only useful as a distinguishing feature if the two species are seen together; also by black facial skin at bill-base encircling eye, and black bill in adult, which lacks yellow patch on bill. **HABITAT, HABITS AND STATUS** Occurs in same areas and feeds in similar way to Eurasian Spoonbill. Globally Endangered due to small world population, habitat loss and pollution, in very limited breeding and wintering areas in East Asia. Hong Kong is one of the most important wintering sites, with around 20 per cent of the total world population (2,693 individuals in 2012 census; BirdLife International, 2013) in Deep Bay area. Formerly Critically Endangered in 1990s, when population was possibly as low as 300 individuals, but since then has increased due to conservation measures. Flagship species for Mai Po NR. (M)

BITTERNS, HERONS & EGRETS

Eurasian Bittern ■ *Botaurus stellaris* 大麻鳽 70–80cm

DESCRIPTION Large, stocky, cryptically coloured, heron-like bird. Rich brown plumage heavily streaked black; paler on underparts. **SIMILAR SPECIES** Immature Black-crowned Night Heron (see opposite) superficially similar but much smaller and darker, lacks any golden tones and has conspicuous whitish spotting on upperwing-coverts. **HABITAT, HABITS AND STATUS** Confined to extensive reed beds. Reclusive. Hunchbacked stance typical, but when alarmed stretches neck and bill vertically and 'freezes'. Uncommon winter visitor and spring migrant to marshes of New Territories, although high counts of up to 31 at Mai Po NR in recent springs suggest site may be an important staging post. (M)

Yellow Bittern
■ *Ixobrychus sinensis* 黃葦鳽 30–40cm

DESCRIPTION Small, buff-coloured bittern. Male has black crown, nape, tail and flight feathers, tawny-buff upperparts and buff underparts, the foreneck and upper breast with darker streaks. Long, pointed bill yellowish-fawn with blackish culmen; yellowish legs. In flight large yellow-buff patch covering inner wing contrasts strikingly with black flight feathers and primary coverts. Female similar to male, but has grey-brown crown and nape, and browner upperparts; streaking on neck and breast is darker and more profuse. Immature heavily streaked brown and buff on both upperparts and paler underparts. **HABITAT, HABITS AND STATUS** Occurs in reed beds, mangroves and dense wetland vegetation, where it is normally hidden from view and best seen when making short flights from one site to another. Uncommon migrant and summer visitor with small numbers overwintering. Largely confined to Deep Bay area, although seen at other wetland sites throughout Hong Kong, particularly on migration. (M, L, I)

■ Bitterns, Herons & Egrets ■

Cinnamon Bittern ■ *Ixobrychus cinnamomeus* 栗葦鳽 40–41cm

DESCRIPTION Small, chestnut-coloured bittern, slightly larger than Yellow Bittern (see opposite). Male has entire upperparts rich cinnamon, and paler underparts and whitish throat. Female duller than male, with dark brown crown and streaking on foreneck and breast. **HABITAT, HABITS AND STATUS** Occurs in wetlands including grassy marshes. Solitary. Predominantly uncommon passage migrant to Deep Bay area; rare in summer; occasionally recorded in winter. (M, L)

Black-crowned Night Heron ■ *Nycticorax nycticorax* 夜鷺 58–65cm

DESCRIPTION Sturdy, 'round-shouldered' heron with striking adult plumage pattern in off-white, grey and black. Crown and back black with dark green gloss; wings, rump and tail grey, and underparts whitish. Two or three long white plumes from crown in summer. Stout bill black, legs yellow and iris deep ruby. Juvenile completely different from adult, resembling much scarcer Eurasian Bittern (see opposite), being a cryptic combination of brown upperparts and lighter brown underparts profusely streaked darker brown with buffish-white spotting on wings. By second winter plumage reflects adult pattern, but colours are muted and brown washed. **HABITAT, HABITS AND STATUS** Wetlands, including vegetated margins of rivers, marshes and mangroves. Crepuscular and nocturnal, roosting in dense vegetation during day. Common resident in Hong Kong wherever suitable habitat exists. Also regular around Hong Kong Harbour. (M, L, H, I)

◾ BITTERNS, HERONS & EGRETS ◾

Striated Heron ◾ *Butorides striata* 綠鷺 40–48cm

DESCRIPTION Small dark heron with shortish neck and legs. Plumage mainly shades of grey, darker with greenish tone on wings where feathers fringed white. Crown to eye level, nape and nuchal crest black. Chin, narrow line down throat and central breast white. Lores and bare skin surrounding eye and legs conspicuously yellow. Bill blackish with lower edge of lower mandible variously shaded pale yellow. Immature browner than adult and lacks head pattern. Neck and sides of breast heavily streaked; underparts off-white streaked brown. **HABITAT, HABITS AND STATUS** Wetlands, particularly mangroves. Tendency to feed at dusk and night-time, although in tidal habitat activity is governed more by tides. Stealthy, inconspicuous species. Uncommon summer visitor largely confined to Deep Bay area; more widespread passage migrant and winter visitor. (M, L, H, I)

Chinese Pond Heron ◾ *Ardeola bacchus* 池鷺 42–52cm

DESCRIPTION Small to medium-sized, short-necked heron with cryptic brown and white plumage in winter. Head, neck, breast and upper back whitish, heavily streaked brown; remainder of upperparts dull buffy-brown. Lower breast and belly white. White wings, hidden at rest, transform appearance in flight. Breeding plumage a handsome combination of rich deep chestnut head, neck and breast, smoky-black upperparts, and plumed scapulars and white underparts. Bill yellow, tipped with black (upper mandible wholly black in winter); legs greenish-yellow to yellow. **HABITAT, HABITS AND STATUS** Wetlands, mainly in northern New Territories, but also widespread elsewhere. (M, L, H, I)

Non-breeding

Breeding

▪ Bitterns, Herons & Egrets ▪

Eastern Cattle Egret
▪ *Bubulcus coromandus* 牛背鷺 46–56cm

DESCRIPTION Small white egret with distinctive stocky, hunchbacked appearance. All white, acquiring orange-buff head, neck, upper breast and mantle in breeding plumage. Short (for an egret) yellow bill and short grey-green to blackish legs respectively turn bright red at base and reddish during courtship period. Juvenile as winter adult but has dark bill. HABITAT, HABITS AND STATUS Throughout much of its geographical range feeds in groups by following cattle, catching disturbed insects, frogs, snakes and lizards. In Hong Kong has also adopted other feeding niches such as banks of fish ponds and landfill sites. Common resident population swelled by passage migrants. Largely confined to northern New Territories, although migrants occur widely elsewhere. An alternative taxonomic treatment considers Eastern Cattle Egret as a subspecies, *B. ibis coromandus*, of Cattle Egret. (M, L, I)

Grey Heron ▪ *Ardea cinerea* 蒼鷺 90–98cm

DESCRIPTION Very large, long-necked, long-legged waterbird, with dagger-like bill, broad, rounded wings and short tail. Mainly grey with whitish crown, neck and underparts. Conspicuous black line from eye stretching back across side of head and, in adults, extending to long plumes falling from rear crown. Bold black streaking down front of neck, and black 'shoulder-patch' on wing bend. Bill yellow; legs and feet ochre. HABITAT, HABITS AND STATUS Wetlands from maritime coasts to reservoirs, fish ponds and marshes. Feeds largely on fish, crustaceans and amphibians captured from statuesque standing position or while walking slowly. Flies strongly with neck retracted on broad and markedly bowed wings. Common winter visitor mainly to Deep Bay area, with some individuals remaining all year and occasionally breeding. (M, L, I)

■ BITTERNS, HERONS & EGRETS ■

Purple Heron ■ *Ardea purpurea* 草鷺 78–90cm

DESCRIPTION Compared with Grey Heron (see p. 27), slightly smaller with a more slender build emphasized by a very long, sinuous neck. Mantle grey, wings purplish-brown, head and neck striped rusty-orange and black, chin and foreneck white. 'Shoulder-patch' and underparts deep chestnut, becoming black on belly and undertail-coverts. Immature duller and browner. **HABITAT, HABITS AND STATUS** Denizen of extensive reed beds; secretive, preferring to remain in deep cover rather than feeding in open as Grey Heron. Most readily seen in flight between reed beds. Uncommon resident and passage migrant largely restricted to Deep Bay area, particularly Mai Po. (M)

Great Egret ■ *Ardea alba* 大白鷺 85–102cm

DESCRIPTION Largest white heron in Hong Kong. Similar in height to Grey Heron (see p. 27), but less robust, with strikingly long, slender neck. Plumage all white with plumes falling from upper breast and, for a short period during courtship, from scapulars. Bill yellow, turning black during courtship; loral skin greenish; legs and feet usually black, becoming red or pinkish-red at height of breeding condition. Juvenile similar to winter adult but lacks plumes. At close range, length of gape line, which extends beyond rear of eye, is diagnostic. **HABITAT, HABITS AND STATUS** Gregarious, often feeding in large numbers along tidal creeks or drained fish ponds, on fish, amphibians and crustaceans and, on land, snakes, lizards, insects and small mammals. Wingbeats more measured than those of other white egrets. Abundant winter visitor and passage migrant predominantly in Deep Bay area. Small breeding colonies in north-east New Territories. (M, L, I)

▪ BITTERNS, HERONS & EGRETS ▪

Intermediate Egret ▪ *Egretta intermedia* 中白鷺 65–72cm

DESCRIPTION Similar to Great Egret (see opposite), but smaller. Plumage entirely white; in breeding plumage has plumes cascading from scapulars and upper breast. Bill proportionately shorter than bills of other white herons, giving somewhat stubby appearance; yellow, often with dark culmen and/or tip, becoming blackish with yellow base during courtship. Bare loral skin yellow, turning yellowish-green at courtship peak. Legs and feet black/blackish. Despite considerable size difference, lone individuals can prove surprisingly difficult to distinguish from Great Egret. **HABITAT, HABITS AND STATUS** Wetlands including estuarine mudflats, preferring low, vegetated areas of freshwater marshes. Habits similar to Great Egret's. Routinely mixes with feeding flocks of other heron species. Present throughout year in Deep Bay area, with regular sightings at Sheun Wan and Starling Inlet, although principally a passage migrant. (M)

Little Egret ▪ *Egretta garzetta* 小白鷺 55–65cm

DESCRIPTION Small to medium-sized, entirely white heron, in breeding plumage with plumes from rear crown (aigrettes), upper breast, mantle and scapulars. Bill black; bare loral skin greenish-grey, becoming orange at courtship peak. Legs black with strikingly yellow feet, a distinctive feature obvious even in flight and at distance. Immature similar, without plumes and with greyish-green feet. **SIMILAR SPECIES** Intermediate Egret (see above) and scarce **Swinhoe's Egret** *E. eulophotes* both distinguished from this species by their yellow bills. **HABITAT, HABITS AND STATUS** Wetlands of varying types with fresh, brackish and salt water, including lakes, rivers, marshes, mangroves and mudflats. Also regular in Hong Kong Harbour. Main food comprises small fish, crustaceans and aquatic insects caught using a variety of techniques, of which 'foot stirring' and 'running', when bird dashes about erratically in shallow water with wings raised, are the most eye-catching. Abundant resident. (M, L, H, I)

■ Bitterns, Herons, Egrets & Cormorant ■

Pacific Reef Heron ■ *Egretta sacra* 岩鷺 58–66cm

DESCRIPTION Dimorphic medium-sized heron. Dark morph entirely slaty-grey except for narrow white stripe down chin and throat. Elongated nape feathers forming small crest and plumes on lower foreneck and scapulars developed in breeding plumage. Stout long bill yellowish-brown; relatively short legs greenish-black to yellowish-green. Immature plumage browner grey than adult's. Pale morph (very rare in Hong Kong) has all-white plumage and is easily confusable with Little Egret (see p. 29) and scarce **Swinhoe's Egret**. **HABITAT, HABITS AND STATUS** Inhabits coastal waters, especially rocky shorelines. Usually a solitary feeder. Uncommon resident mainly on south coast of Hong Kong Island and outlying islands. (I)

Great Cormorant ■ *Phalacrocorax carbo* 普通鸕鶿 80–100cm

DESCRIPTION Large, long-necked black seabird with elongated body, longish tail and long, relatively slender, distinctly hooked bill. Sexes are similar; breeding, non-breeding and immature plumages distinct. Adult in winter black except for white face- and throat-patch. Adult in summer plumage (attained in Hong Kong by some individuals) similar, but acquires glossy green sheen, white thigh-patch, and white plumes from crown to lower neck. Immature dull, with dark brown upperparts, and pale dusky or whitish underparts. **HABITAT, HABITS AND STATUS** Aquatic, in both coastal marine and inland waters. Flight strong and direct; often in large flocks that form a line or V formation. Swims low in water, diving frequently to feed on fish and, when perched, often holds wings outstretched in 'spread-eagle' position. Stands upright on short legs set towards rear of body. Awkward, waddling gait. Abundant winter visitor mainly to Deep Bay area. (M)

Roost

▪ Osprey, Kites, Hawks & Eagles ▪

Western Osprey ▪ *Pandion haliaetus* 鶚 55–58cm

DESCRIPTION Medium-sized brown and white raptor. White head, brown streaking on crown and broad black eye-stripe. Upperparts dark brown; underparts white with breast-band of rusty-buff streaks. In flight rather long, narrow wings and shortish tail produce distinctive angular outline. Viewed from below white body and wing-coverts, grey flight feathers with black-tipped primaries and conspicuous black carpal-patch diagnostic. Sexes and juvenile alike. **HABITAT, HABITS AND STATUS** Confined to aquatic habitats, both freshwater and sea water, by specialized live fish diet. Catches fish in dramatic aerial dive, entering water feet-first, then flies to exposed perch (often a post) to feed and rest. Regularly employs cumbersome hover preparatory to dive. Common winter visitor mainly to Deep Bay area. (M)

Black-winged Kite ▪ *Elanus caeruleus* 黑翅鳶 31–35cm

DESCRIPTION Unmistakable, small pale raptor. Upperparts silvery-grey; underparts paler, almost white. Bold black 'shoulder', underwing-tips and small patch around bright, ruby eye. Distinctly broad-based wings. Immature similar to adult, with brownish wash on crown, mantle and breast, and pale tips to upperwing-coverts. Iris dull red. **HABITAT, HABITS AND STATUS** Favours open ground with scattered trees. Frequently hovers when hunting. Solitary. Uncommon year-round visitor with majority of records from Deep Bay area. (M, L)

KITES, HAWKS & EAGLES

Crested Honey Buzzard
■ *Pernis ptilorhyncus* 鳳頭蜂 55–65cm

DESCRIPTION Fairly large, small-billed, small-headed, long-necked and long-tailed raptor with short, inconspicuous crest. Plumage variable. Typical male has greyish-brown upperparts, grey 'face' and whitish throat with narrow dark border line and mesial stripe. Underparts creamy, barred brown to rufous. Long, broad, rounded wings pinched in at bases. Undersides rather pale with conspicuous dark trailing edge; three bars across primaries and secondaries, and dark tail with wide pale central band. Female similar to male, with brown 'face', four bars on remiges and pale tail narrowly barred blackish-brown. Immature typically has paler underside with less distinct wing-bars. **HABITAT, HABITS AND STATUS** Active flight comprises several wingbeats, then longish glide on horizontal wings. Uncommon passage migrant mostly in autumn, with occasional winter records. (M, T, I)

Crested Serpent Eagle
■ *Spilornis cheela* 蛇鵰 56–74cm

DESCRIPTION Large raptor with rather short, broad, rounded wings, large, flat head, short, bushy crest bulging from nape and bright yellow loral skin. Upperparts dark brown; underparts paler ruddy-brown. Small, well-defined white spots on coverts, lower breast and belly. Striking underwing pattern of warm brown body and coverts, the latter bordered by black band, wider white band and black trailing edge. Tail black with bold white central band. Juvenile much paler than adult, with dark cream-streaked underparts and upperparts mottled brown. Tail banded brown and white. Voice: single, double or triple, high-pitched, shrieking whistle, the last with a slight pause between the second and lower pitched third syllables. **HABITAT, HABITS AND STATUS** Woodland specialist feeding mainly on reptiles, especially snakes. Soars for long periods and is very vocal, especially in spring, when loud, far-carrying calls draw attention to its presence. Uncommon local resident. (M, L, T, I)

KITES, HAWKS & EAGLES

Greater Spotted Eagle
■ *Clanga clanga* 蛇鵰 65–72cm

DESCRIPTION Bulky, broad-winged, medium-sized eagle with short tail. Adult essentially blackish-brown, although upperwing often shows white patch at base of primaries, and uppertail-coverts are often white. Prominent yellow cere contrasts with dark head. Immature copiously spotted on upperwing-coverts, with white tips of greater coverts forming distinct pale bar. Trailing edge of wings, uppertail-coverts and tip of tail white. Undersides of wings at all ages have dark coverts and marginally lighter flight feathers. **HABITAT, HABITS AND STATUS** Uncommon winter visitor almost exclusively to Deep Bay area. (M)

Eastern Imperial Eagle ■ *Aquila heliaca* 白肩鵰 68–84cm

DESCRIPTION Large, majestic eagle. Essentially dark brown with tawny-gold crown and hindneck, and two-toned tail with pale base and wide dark terminal band. Small white patch on scapulars. Sexes are similar. Juvenile buffish-brown. In flight viewed from above primary and greater coverts, rear edges of wings and tail tipped white; rump off-white. From below shows pale streaked body and coverts; blackish outer primaries with black 'fingers', secondaries and tail, and obviously pale inner primaries forming a wedge-shaped patch. Trailing edges of wings and tail-tip white. **HABITAT, HABITS AND STATUS** Open country. Tends to hold tail tightly closed even when soaring. Uncommon but regular winter visitor to Deep Bay area. (M)

Juvenile

KITES, HAWKS & EAGLES

Bonelli's Eagle ■ *Aquila fasciata* 白腹隼鵰 68–84cm

DESCRIPTION Medium-sized eagle with rather long tail. Upperparts grey-toned brown with white patch on mantle, and grey tail with broad, dark terminal band. Breast and belly white with dark streaking. Underwing dominated by wide, dark, diagonal covert-band separating pale forearm from darkish remiges. Juvenile lacks mantle-patch; diagonal covert-band barely apparent or absent; pale rufous underparts and underwing-coverts. **HABITAT, HABITS AND STATUS** Tends to hold tail closed even when soaring. Often seen in pairs following same route each day. Uncommon resident over open land and hillsides in New Territories and Lantau Island. (M, L, T, I)

Crested Goshawk ■ *Accipiter trivirgatus* 鳳頭鷹 ♂ 30–38cm, ♀ 39–46cm

DESCRIPTION Largest *Accipiter* occurring regularly in Hong Kong. Male has grey head, inconspicuous nuchal crest and brownish-grey upperparts. Underparts white and throat with prominent black mesial stripe; upper breast with wide rufous-brown streaks, below which rufous-brown bands reach to white undertail-coverts. Tail divided into broad black and brown bands (4–5 of each) of approximately equal width. Female similar to male, though larger and browner. Juvenile has creamy underparts, heavily blackish streaked breast, large spots on belly and barred thighs. **HABITAT, HABITS AND STATUS** Deciduous woodland, *fung shui* woods. Common and widespread throughout Hong Kong; most readily seen in spring, when soaring and winnowing display flight is performed for lengthy periods. (M, L, T, H, I)

▪ Kites, Hawks & Eagles ▪

Chinese Sparrowhawk
▪ *Accipiter soloensis* 赤腹鷹 27–35cm

DESCRIPTION Medium-sized *Accipiter* with comparatively short tail. Male noticeably pale with unmarked light ash-grey head and upperparts, and darker grey wings with black primaries. Underparts white; breast washed pink without streaking or barring. In flight wings appear rather pointed; underwing strikingly pale with buff coverts, white flight feathers, the primaries prominently tipped black and 4–6 narrow, incomplete dark bars on tail. Female similar to male but slightly larger, with grey throat, orange-buff breast and yellow iris (dark brown to red in adult male). Juvenile dark brown above, and white below with rufous-brown streaking on breast and barring on flanks. **HABITAT, HABITS AND STATUS** Common passage migrant, most frequent in spring when large flocks are occasionally recorded. (M, L, T, I)

Japanese Sparrowhawk ▪ *Accipiter gularis* 日本松雀鷹 ♂ 23–26cm, ♀ 26–30cm

DESCRIPTION Smallest *Accipiter* regularly recorded in Hong Kong. Male has slate-grey head and upperparts, and 4–5 medium-width bands on tail. White throat with faint dark mesial stripe. Breast and flanks brownish-orange, barred on belly sides and 'trousers'; white central underparts and undertail-coverts. Tail appears relatively short. Female larger than male; upperparts greyish-brown; underparts barred brown and white; more pronounced dark mesial and yellow iris (red in male). Juvenile dark brown above; large teardrops on breast and dark mesial. **SIMILAR SPECIES** Besra (see p. 36) has shorter wings and more strongly marked underparts, including more prominent mesial and broader tail-bands. **HABITAT, HABITS AND STATUS** Forest bird that adopts more open areas in winter. Uncommon but widespread passage migrant, most frequent in autumn, with occasional winter records. (M, L, T, I)

KITES, HAWKS & EAGLES

Besra ■ *Accipiter virgatus* 松雀鷹 ♂ 24–30cm, ♀ 31–36cm

DESCRIPTION Small sparrowhawk. Male has dark grey upperparts, boldly streaked upper breast, and broadly barred lower breast and belly in deep rufous-brown and white, with orange-buff wash to breast sides. Conspicuous dark mesial. Banded tail comprises 4–6 wide blackish bars alternating with slightly narrower pale grey bars. Iris red. Female larger than male; darker and browner with yellow iris. Juvenile has similar upperparts to female's, and underparts with large blackish teardrop spotting. **HABITAT, HABITS AND STATUS** Open and mature woodland. Common and widespread resident and migrant. (M, L, T, H, I)

Eastern Marsh Harrier ■ *Circus spilonotus* 白腹鷂 43–54cm

DESCRIPTION Medium-sized raptor with rather long, broad wings and long tail. Male has mainly black head and upperparts, the latter streaked white; secondaries and tail sooty-grey; underparts white; throat and upper breast strongly streaked black. In flight, underwing white with black primary tips. Rump-patch white. Female dark brown with rufous underparts. Head and breast mottled creamy-buff, tail with narrow dark bands and rump-patch varying from white to buff or absent. Juvenile variable, but essentially has dark brown upperparts, creamy crown, and neck, throat and upper breast all variably streaked brown. Lower breast and belly rufous-brown. **HABITAT, HABITS AND STATUS** Wetlands, particularly those that are well vegetated, for example extensive reed beds. Distinctive hunting technique, quartering low over the ground in slow, wavering flight with wings held in shallow V. Common winter visitor mainly to Deep Bay area. (M)

Immature male

▪ KITES, HAWKS & EAGLES ▪

Pied Harrier ▪ *Circus melanoleucos* 鵲鷂 41–49cm

DESCRIPTION Smallish, slender harrier. Male strikingly black, grey and white, with black head, mantle and upper breast, white 'shoulder-patch', grey wings and tail, and white lower breast and belly. In flight appears grey with black head, mantle, outer primaries and median coverts, and white forearms. Female's upperparts mottled brown and black; greyish upperwing with pale forearm, white uppertail-coverts and grey tail barred black. Underparts pale with dark streaking on upper breast. Juvenile resembles female but darker brown; uppertail-coverts buffish and lacks pale forearm. **HABITAT, HABITS AND STATUS** Uncommon migrant, mostly in autumn, to Deep Bay area. (M, L)

Juvenile

Black Kite ▪ *Milvus migrans* 黑鳶 61–66cm

DESCRIPTION Medium-sized dark brown raptor with longish, parallel-sided wings and long, notched tail that is diagnostic. Upperparts show paler coverts and underparts have a distinct pale patch at base of primaries. Immature has heavily streaked underparts, and white tips to coverts forming narrow band across upperwings and underwings. **HABITAT, HABITS AND STATUS** By far the most numerous and conspicuous raptor in Hong Kong. Widespread due to its readiness to scavenge in all but the most wooded areas. Favours urban sprawls as much as uninhabited areas and is almost always present, soaring over Hong Kong Harbour and Hong Kong Island. Forms large winter roosts of up to 1,000 birds, for example at Magazine Gap. Abundant resident and winter visitor. (M, L, T, H, I)

▪ KITES, HAWKS & EAGLES ▪

White-bellied Sea Eagle ▪ *Haliaeetus leucogaster* 白腹海鵰 69–92cm

DESCRIPTION Large fish eagle. Head, underparts and tail white. Upperparts dark grey. In flight head, upperparts, wedge-shaped tail (except for black basal band), and wing-coverts white and flight feathers black. Seen from above, pale head and neck contrast with largely grey upperparts and black primaries. Tail has wide white terminal band. Sexes are similar, but female is larger than male. Immature, when perched, appears mainly streaked brown with dark brown upperparts, pale buffish head and dusky brown underparts. In flight underparts distinctly patterned, with brownish body and underwing-coverts bordered by pale and dark bands, grey-brown secondaries, black primary tips and a whitish base to primaries forming a pale panel. Tail whitish shading to brown at tip. **HABITAT, HABITS AND STATUS** Sea coasts and estuaries. Perches conspicuously on bare branches, boulders and similar structures, adopting noticeably upright stance. Soars with wings held in deep V. Snatches food items from surface of water. Uncommon, widespread resident, mainly around coast and offshore islands. (M, I)

Grey-faced Buzzard ▪ *Butastur indicus* 灰臉鵟鷹 41–48cm

DESCRIPTION Medium-sized raptor with longish wings and tail, giving somewhat *Accipiter*-like jizz and distinguishing it from Eastern Buzzard (see opposite). Upperparts dark brown; head shows grey cheeks, pale supercilium when seen well, and white throat with black mesial stripe. Underparts warm brown merging to barred brown and white lower breast and belly. Sexes are similar. Juvenile has pale underparts heavily streaked blackish. **HABITAT, HABITS AND STATUS** Uncommon passage migrant in spring, when in some years flocks of more than 100 birds are recorded. (M, I)

KITES, HAWKS, EAGLES & CRAKE

Eastern Buzzard
■ *Buteo japonicus* 普通鵟 40–52cm

DESCRIPTION Medium-sized, thickset raptor with longish wings and rather short, rounded tail. Notable for its wide plumage variation. Most often shows plain dark upperparts, and pale underparts with dark flanks and chest-band. In flight underparts appear relatively plain, with dark 'fingers', trailing edges of wings and carpal-patches, the last being diagnostic. Tail usually fairly plain or finely barred medium brown, although colour varies from creamy to pale rufous. **HABITAT, HABITS AND STATUS** Open country associated with wooded areas. When foraging occasionally hovers. Common and widespread winter visitor. (M, L, T, I)

Slaty-legged Crake ■ *Rallina eurizonoides* 灰腳秧雞 21–25cm

DESCRIPTION Medium-sized crake with deep orange-brown head, neck and breast. Wings, upperparts and tail dark brown. Belly and undertail-coverts banded black and white. Legs greenish-grey to grey. Presence revealed by distinctive monotonous, two-note call repeated at regular intervals during dusk and night. **HABITAT, HABITS AND STATUS** Unlike other crakes, occurs in widespread, well-wooded areas. Skulking; walks stealthily on woodland floor. Locally common summer visitor and migrant, and scarce winter visitor. (T, I)

CRAKES & RAILS

White-breasted Waterhen ■ *Amaurornis phoenicurus*
白胸苦惡鳥 28–33cm

DESCRIPTION Large rail with unusual and striking plumage colouration. Cap, hindneck and upperparts greyish-black. Cheeks, foreneck, breast and upper belly white. Rear flanks and lower belly orange-rufous, merging into deeper toned undertail-coverts. Iris deep red; bill yellow with red base to upper mandible; legs and feet ochre. Sexes are similar. Immature duller than adult, with brownish tones to upperparts and brown iris. Voice: very vocal, including wide repertoire of squawks and bubbling calls as well as continuously repeated, two-note call. **HABITAT, HABITS AND STATUS** Common and widespread resident of all types of wetland. Walks slowly while foraging, often in the open, running swiftly to cover if disturbed. (M, L, T, H, I)

Ruddy-breasted Crake ■ *Porzana fusca* 紅胸田雞 21–23cm

DESCRIPTION Medium-sized crake with brown upperparts and deep brownish-chestnut underparts, small whitish throat-patch, and black, narrowly barred white rear belly, flanks and undertail-coverts. Bill dark grey; iris and legs red. **SIMILAR SPECIES** Slaty-legged Crake (see p. 39) has entire head, mantle, throat and breast bright orange-brown; belly and undertail-coverts boldly striped black and white; legs greenish-grey to blackish. Vagrant **Band-bellied Crake** *P. paykullii* has bold black and white stripes across belly to undertail-coverts. **HABITAT, HABITS AND STATUS** Stealthy inhabitant of thick wetland vegetation. Uncommon passage migrant and winter visitor principally to Deep Bay area. (M, L)

▪ Gallinule & Coot ▪

Common Moorhen ▪ *Gallinula chloropus* 黑水雞 32–35cm

DESCRIPTION Medium-sized rail. Head, neck and underparts black; upperparts dark brown. White line along flanks and outside edge of undertail-coverts. Striking red iris, frontal shield and bill, the last tipped bright yellow. Legs and feet greenish-yellow. Sexes are similar. Immature duller than adults, with mainly brown upperparts and dark grey underparts. **HABITAT, HABITS AND STATUS** Freshwater and brackish wetlands of all kinds, for example salt-marsh channels and margins of *gei wai* with emergent cover. Walks with high-stepping, hesitant strides, characteristically with tail raised and jerked constantly. Also jerks head when swimming. Common winter visitor, resident and migrant mainly in northern New Territories. (M, L)

Eurasian Coot ▪ *Fulica atra* 骨頂雞 36–38cm

DESCRIPTION Large, stout rail with black head and neck, and rest of plumage grey-black. White bill and frontal shield, red iris, yellowish tibia and grey-green tarsus. **HABITAT, HABITS AND STATUS** Wetlands, preferring open, shallow waters fringed with vegetation. Not at all reclusive, spending much time in open water and congregating in flocks. Dives to feed. Uncommon winter visitor, mainly to Deep Bay area. (M)

▪ STILT & AVOCET ▪

Black-winged Stilt ▪ *Himantopus himantopus* 黑翅長腳鷸 35–40cm

DESCRIPTION Supremely elegant black and white wader with extraordinarily long legs. Black mantle and wings; remainder of plumage white except for dusky rear crown and hindneck, more extensive in breeding female. Body size similar to Common Redshank's (see p. 52), with pinkish-red legs and fine black bill approximately doubling total measurement. **HABITAT, HABITS AND STATUS** Shallow fresh or brackish waters, mainly in Deep Bay area. Common passage migrant and winter visitor with small numbers breeding. (M, L)

Pied Avocet ▪ *Recurvirostra avosetta* 反嘴鷸 42–45cm

DESCRIPTION Elegant black and white wader of medium size, made to appear larger by long, blue-grey legs and unique long, upcurved black bill. Essentially white with black cap to below eye, and black hindneck, sides of mantle, wing-coverts stripe and outer primaries. Sexes are similar. Juvenile has brown markings on mantle and scapulars. **HABITAT, HABITS AND STATUS** Shallow saline waters including mudflats. Distinctive feeding technique consisting of continuous scything sweeps of head and bill through water. Swims confidently, regularly gathering in flocks to rest on tidal waters for lengthy periods. Abundant winter visitor restricted to Deep Bay area. (M, L)

◾ Plovers ◾

Grey-headed Lapwing ◾
Vanellus cinereus 灰頭麥雞 34–37cm

DESCRIPTION Large, stocky wader with grey head and breast, light brown upperparts, black breast-band and white belly. Bright, deep yellow bill and long legs, the former with an obvious black tip. Iris red, eye-ring and small wattles at base of bill deep yellow. In flight light brown coverts, white secondaries and black primaries create strikingly bold pattern. Tail white, broadly tipped with black. **HABITAT, HABITS AND STATUS** Grassy and wetland areas. Uncommon winter visitor to north-west New Territories, particularly Kam Tin. (M, L)

Pacific Golden Plover ◾ *Pluvialis fulva* 太平洋金斑鴴 22–26cm

DESCRIPTION Medium-sized plover with yellow, black and white-spangled upperparts and dingy, mottled brownish-grey underparts. Transformed in breeding plumage, which some spring individuals begin to acquire in Hong Kong, with brighter, more intensely coloured upperparts and jet-black underparts separated by bright white band running from forehead across head, neck and sides of breast, and onto flanks. **HABITAT, HABITS AND STATUS** Gregarious. Common passage migrant and winter visitor to Hong Kong, where it feeds on intertidal mudflats in Deep Bay. (M)

◾ Plovers ◾

Grey Plover ◾ *Pluvialis squatarola* 灰斑鴴 27–30cm

DESCRIPTION Large, rather heavy-looking plover. In non-breeding plumage has distinctly plain appearance, with brownish-grey upperparts, black primaries and dusky underparts. In flight black axillaries are diagnostic. Breeding plumage, which some individuals begin to acquire in Hong Kong, is striking combination of silver, white and black-spangled upperparts and jet-black underparts. Sexes are similar. **HABITAT, HABITS AND STATUS** Feeds on intertidal mudflats, adopting run-stop-peck technique characteristic of plovers, with the stop often held motionless for many seconds. Abundant winter visitor, almost exclusively to Deep Bay area. (M)

Little Ringed Plover ◾ *Charadrius dubius* 灰斑鴴 14–15cm

DESCRIPTION Small plover with brown upperparts and white underparts. In breeding plumage has brown cap; white frontal bar enclosed by two black bars, one across forehead the other from bill-base, joining in front of eye and extending across 'face'; distinct yellow eye-ring; white throat and neck-collar, and broad black breast-band. Bill black with small yellow spot at base of lower mandible; dull pinkish, occasionally yellowish legs. Winter plumage similar, but duller and without forehead-bars. Immature plainer than adult, with brown head and breast-band, and pale fringes to upperparts feathers. **SIMILAR SPECIES** Kentish Plover (see opposite) shows white wing-bar in flight, and breast-band reduced to patch on either side of breast. Rare **Common Ringed Plover** *C. hiaticula* also has white wing-bar, and less prominent eye-ring. **HABITAT, HABITS AND STATUS** Dryer areas of freshwater marshes and fish ponds. Shuns intertidal mudflats. Ground-loving species; runs quickly and ranges widely when foraging. Usually flies fairly low to the ground. Common winter visitor and passage migrant with small breeding population, mainly in north-west New Territories. (M, L, I)

Non-breeding    *Adult breeding*

■ PLOVERS ■

Kentish Plover ■ *Charadrius alexandrinus* 環頸鴴 15–17cm

DESCRIPTION Superficially similar to Little Ringed Plover (see opposite). Male in non-breeding plumage has brown upperparts and white underparts. Brown cap, wide blackish-brown eye-stripe, and white forehead, supercilium and hind-collar. Brown patch protrudes from shoulder onto sides of breast. In breeding plumage cap takes on tawny hue, eye-stripe and breast-patch become black, and black patch separates cap from white forehead. Upperparts brighter brown. Female similar to male, but lacks chestnut cap and black frontal bar. Eye-stripe and breast-patch blackish-brown. Bill and legs black in all plumages. **SIMILAR SPECIES** See Little Ringed Plover. **HABITAT, HABITS AND STATUS** Gregarious. Abundant winter visitor mainly to mudflats of Deep Bay. (M)

Non-breeding

Male breeding

Lesser Sand Plover ■ *Charadrius mongolus* 蒙古沙鴴 19–21cm

DESCRIPTION Smallish plover with rather long grey legs. Essentially plain in non-breeding plumage, with dull brown upperparts and pale underparts relieved by distinct pale supercilium and dull grey-brown breast-band. Much brighter in breeding plumage, with warm tone to upperparts and broad, brick-red breast-band, shading onto nape and crown, the upper border edged with narrow black line. Forehead, chin and throat white; black mask extending to lores and separately across forehead. Sexes are similar. Immature similar to winter adult. **SIMILAR SPECIES** See Greater Sand Plover (see p. 46). **HABITAT, HABITS AND STATUS** Regularly mixes with Greater Sand Plovers in Hong Kong, where it is uncommon on spring passage and even less numerous in autumn and winter. Mainly frequents intertidal mudflats of Deep Bay. (M)

Breeding

Non-breeding

▪ Plovers & Painted-snipe ▪

Greater Sand Plover ▪ *Charadrius leschenaultii* 鐵嘴沙鴴 22–25cm

DESCRIPTION Larger than Lesser Sand Plover (see p. 45), although size difference is not always obvious in lone individuals. Plumage patterns in all seasons and at all ages similar to that species, but rufous parts more orangey and less red, upperparts paler brown, bill longer with more bulbous tip, and legs longer and paler, the colour varying from yellowish to pinkish. Structure less refined and more gangly than that of Lesser Sand Plover. In flight feet protrude beyond tip of tail. **HABITAT, HABITS AND STATUS** Habitat and habits similar to Lesser Sand Plover's. Abundant passage migrant, with small numbers over-wintering, almost exclusively in Deep Bay area. (M)

Breeding

Non-breeding

Greater Painted-snipe ▪ *Rostratula benghalensis* 彩鷸 23–26cm

DESCRIPTION Unusually patterned and coloured, rather bulky, medium-sized wader. Female much brighter than male, with deep red-brown head and neck, golden-buff central crown-stripe and conspicuous white patch stretching back from eye. Breast blackish, bordered by striking white band extending to mantle. Upperparts largely grey-green, finely barred black. Underparts creamy-white; tail mainly grey with black bars and golden-buff spots. Longish, heavy bill droops noticeably. Male has olive-brown head and neck, the lower 'face' and breast mottled white. Eye-patch and breast-band buff and upperparts dominated by large, roundish buff spots. Immature similar to adult male. **HABITAT, HABITS AND STATUS** Male incubates and rears young. Crepuscular and nocturnal. When flushed, broad, rounded wings are apparent; often trails legs in flight and quickly drops into cover. Uncommon resident in well-vegetated marsh and wet agricultural areas of north-west New Territories. (M, L)

▪ Jacana & Woodcock ▪

Pheasant-tailed Jacana
▪ *Hydrophasianus chirurgus* 水雉 39–58cm; *c.* 31cm in non-breeding plumage

DESCRIPTION Exotically plumaged jacana with long, downward-curving tail in breeding plumage. Adult has white head, throat and foreneck; yellow rear crown and hindneck, with narrow vertical black stripe separating white from yellow. Body black to blackish-brown, with large white wing-patch. Long grey legs and remarkably extended toes. Moults long tail in winter, when plumage is duller. Crown, rear neck and mantle brown; black neck-stripe wider, extending across breast and face as an eye-stripe. Wide buff supercilium. Underparts white. Immature resembles winter adult. **HABITAT, HABITS AND STATUS** Long toes an adaption to aid walking on floating vegetation. Swims well. Uncommon migrant to freshwater marshes, principally at Mai Po, Lok Ma Chau and Long Valley. (M, L)

Eurasian Woodcock ▪ *Scolopax rusticola* 丘鷸 33–35cm

DESCRIPTION Plump, rotund, medium-sized, cryptically coloured wader of woodland. Barred plumage a mix of rufous, buff, brown and black. Crown broadly barred black, eye large and bill long and straight. In flight broad, rounded wings and larger size distinguish it from snipe species. **HABITAT, HABITS AND STATUS** Largely nocturnal, feeding on damp ground in and on edges of woodland. Sits tight and is rarely seen unless flushed. Uncommon but widespread autumn passage migrant and winter visitor. (L, T, I)

◾ SANDPIPERS & SNIPES ◾

Pintail Snipe ◾ *Gallinago stenura* 針尾沙錐 25–27cm

DESCRIPTION Medium-sized, solid and rather rotund, with cryptic brown plumage, long, straight bill and greenish legs. SIMILAR SPECIES Doubtfully separable in the field from scarce **Swinhoe's Snipe**. See Common Snipe (below) for distinguishing features from that species. HABITAT, HABITS AND STATUS Inhabits freshwater marshes, fish ponds and wet agricultural land mainly in northern New Territories. Unobtrusive, crouching when alarmed. Common passage migrant with smaller numbers over-wintering. (M, L)

Common Snipe

◾ *Gallinago gallinago* 扇尾沙錐 25–27cm

DESCRIPTION Medium-sized wetland bird with long, straight bill, dumpy body and cryptic brown plumage except for white belly. Head has two black parallel crown-stripes, and black lines through eye and across cheek. Legs relatively short and greenish. Very similar to Pintail Snipe (see above) and **Swinhoe's Snipe** *G. megala*, and best distinguished in flight by distinct white trailing edges to secondaries and whitish, not dark-barred underwing. Also distinguishable (with experience) from Pintail Snipe by slightly longer bill, higher pitched, rasping call, and distinctly longer projection of tail beyond primary tips. HABITAT, HABITS AND STATUS Sits tight on the ground, but when flushed escape flight is fast, steep and erratic. Common passage migrant and winter visitor mainly to Mai Po NR, as well as wet meadows and agricultural land elsewhere in New Territories. (M, L, I)

▪ Sandpipers & Snipes ▪

Asian Dowitcher
▪ *Limnodromus semipalmatus* 半蹼鷸 33–36cm

DESCRIPTION Reminiscent of a godwit, although smaller and with shorter neck and stouter, all-black bill. In winter upperparts and breast brownish-grey, with paler fringes to mantle and wing feathers. Distinct pale supercilium and dark loral line. Belly and vent white. In breeding plumage head and most of underparts bright chestnut, with white lower belly and dark chevrons on flanks. Upperparts brownish-black fringed chestnut and buff or white. **HABITAT, HABITS AND STATUS** Common spring migrant, but scarce in autumn, only on intertidal mudflats of Deep Bay and at Mai Po NR. Feeds by vigorously probing in mud. (M)

Black-tailed Godwit ▪ *Limosa limosa* 黑尾塍鷸 36–43cm

DESCRIPTION Largish, rather elegant wader, with long legs and long, straight bill. In winter plumage is plain with grey upperparts, head and neck, and white underparts. In breeding plumage acquires brownish-orange head, neck and breast, the last and upper belly barred black. Upperparts chequered black and brownish-orange. Bill dull pink with dark tip. Flight pattern unique among waders of region, with bold white wing-bar contrasting with black flight feathers and outer coverts, and white rump and uppertail-coverts contrasting with black tail. **SIMILAR SPECIES** See Bar-tailed Godwit (see p. 50). **HABITAT, HABITS AND STATUS** Gregarious. Abundant winter visitor and passage migrant almost exclusively to Deep Bay area. (M)

■ SANDPIPERS & SNIPES ■

Bar-tailed Godwit ■ *Limosa lapponica* 斑尾塍鷸 38–46cm

DESCRIPTION Very similar to Black-tailed Godwit (see p. 49), but a little larger, with longer, slightly upturned bill and shorter legs. In winter plumage has darker centres to upperparts feathers, giving mottled appearance. Bill dark with flesh base. In flight pale wedge on upper tail, rump and back and absence of white wing-bar make separation straightforward. In breeding plumage whole of underparts russet, but beware of moulting individuals that have paler lower belly similar to Black-tailed Godwit's. **HABITAT, HABITS AND STATUS** Gregarious. Uncommon passage migrant and irregular winter visitor mostly to Deep Bay intertidal mudflats. (M)

Whimbrel ■ *Numenius phaeopus* 中杓鷸 40–42cm

DESCRIPTION Large, long-legged wader with long, downcurved bill. Rather uniform grey-brown plumage; upperparts feathers with darker centres; paler underparts streaked brown on breast, with arrow-shaped marks on flanks. Five-striped head pattern distinctive, with pale crown-stripe bordered by dark stripe and below a pale supercilium. Voice: flight call a repetitive (usually seven-note), fairly high-pitched, rippling *di-di-di-di-di-di-di*. **SIMILAR SPECIES** Both Eurasian Curlew and Far Eastern Curlew (see opposite) are larger, with significantly longer bills and no head stripes. Latter also lacks pale wedge on lower back, rump and uppertail-coverts, which is so conspicuous in flight on Eurasian Curlew. **HABITAT, HABITS AND STATUS** Common passage migrant and scarce winter visitor mainly to intertidal mudflats of Deep Bay, but also occasionally along sea coasts on passage. (M)

■ Sandpipers & Snipes ■

Eurasian Curlew
■ *Numenius arquata* 白腰杓鷸 50–60cm

DESCRIPTION Large, all mid-brown wader with improbably long, strongly decurved bill. In flight, shows white wedge on lower back and rump, a feature that distinguishes it from similar Far Eastern Curlew (see below). Voice: call a ringing, melodic *coor-lee*. **SIMILAR SPECIES** See Far Eastern Curlew and Whimbrel (opposite). **HABITAT, HABITS AND STATUS** Gregarious and vocal. Abundant winter visitor largely confined to intertidal mudflats of Deep Bay. (M)

Far Eastern Curlew ■ *Numenius madagascariensis* 紅腰杓鷸 53–66cm

DESCRIPTION The largest curlew, although size is not helpful in identification of lone individuals. Very similar to Eurasian Curlew (see above), from which it is easily distinguished in flight by its lack of pale wedge on rump and lower back. Even longer billed than Eurasian Curlew, and plumage is generally brighter toned. Underwing barred brown (mainly white in Eurasian). Voice: call similar to Eurasian Curlew's, but lower pitched. **SIMILAR SPECIES** See Eurasian Curlew and Whimbrel (opposite). **HABITAT, HABITS AND STATUS** Uncommon passage migrant mainly in spring to intertidal mudflats of Deep Bay. (M)

▪ Sandpipers & Snipes ▪

Spotted Redshank ▪ *Tringa erythropus* 鶴鷸 29–31cm

DESCRIPTION Longish bill and legs add to refined appearance of this medium-sized wader. In non-breeding plumage essentially pale grey with paler underparts. In flight shows conspicuous white patch from upper rump to back. Breeding plumage unique among sandpipers, being entirely smoky-black except for white patch on back and small white spots on body. Bill fine, drooping at extreme tip, and black with red base to lower mandible. Legs dark red, becoming lighter in winter. Voice: distinctive, sharp *chew-it* call. **HABITAT, HABITS AND STATUS** Flooded grassland, fish ponds and mudflats. When numerous, forms significant flocks and regularly swims when feeding. Common spring passage migrant, less numerous in autumn and winter, mainly to Deep Bay area. (M, L)

Common Redshank ▪ *Tringa totanus* 紅腳鷸 27–29cm

DESCRIPTION Medium-sized wader with pale grey-brown upperparts and breast, and white belly. Longish legs and bill bright red, the latter dark tipped. In flight bold upperparts' pattern is revealed, showing broad white secondaries and inner primary tips, and white patch on upper rump and back. Voice: often first species to raise alarm with a loud, urgent, piping *tlee, tlee, tlee, tle* and distinctive downward-inflecting *teu-hu-hu*. **HABITAT, HABITS AND STATUS** Alert and noisy. Abundant passage migrant and winter visitor to intertidal mudflats of Deep Bay and adjacent wetlands, especially Mai Po NR. (M)

▪ SANDPIPERS & SNIPES ▪

Marsh Sandpiper ▪ *Tringa stagnatilis* 澤鷸 22–24cm

DESCRIPTION Slim, long-legged, elegant sandpiper with extremely fine, longish bill. In winter plumage crown, hindneck, eye-stripe and upperparts are grey, wing-coverts often darker; fairly distinct white supercilium; face and underparts white. Legs vary from dull green to distinctly yellow. In breeding plumage upperparts rich brown, copiously spotted black; neck and breast heavily streaked black. In flight shows white wedge on upper rump and lower back; legs and feet protrude well beyond tail. **SIMILAR SPECIES** See Common Greenshank (below). **HABITAT, HABITS AND STATUS** Most numerous in spring, but also common autumn and winter visitor to intertidal mudflats of Deep Bay and adjacent wetlands, especially Mai Po NR. (M, L)

Common Greenshank ▪ *Tringa nebularia* 青腳鷸 30–33cm

DESCRIPTION Largest *Tringa* in the region, with long, heavy, slightly upturned bill and grey-green legs. In winter plumage rather plain with grey upperparts and white underparts. Head, hindneck and sides of upper breast finely streaked. In flight shows white wedge on upper tail, rump and lower back. In breeding plumage streaking heavier and extending to upper breast. Voice: flight call a far-carrying, ringing *tu-tu-tu*. **SIMILAR SPECIES** Marsh Sandpiper (see above) is smaller and slimmer with a fine, straight bill, and Nordmann's Greenshank (see p. 54) has shorter legs and a thicker based bill. **HABITAT, HABITS AND STATUS** Abundant winter visitor and passage migrant (less numerous in autumn), mainly to intertidal mudflats of Deep Bay and adjacent wetlands. (M, L)

◾ SANDPIPERS & SNIPES ◾

Nordmann's Greenshank ◾ *Tringa guttifer* 小青腳鷸 29–32cm

DESCRIPTION Very similar to Common Greenshank (see p. 53), but slightly smaller and stockier, with shorter, distinctly yellowish legs and thicker based bill. On close view small palmations between front three toes are diagnostic (Common Greenshank has small palmation between inner and middle toe only). In winter plumage upperparts fringed white and underparts plain without streaking. In breeding plumage upperparts have dark centres, producing a boldly spangled effect, and breast shows obvious peppering of small black spots. Voice: flight call a deep-throated *gwaark*, quite different from Common Greenshank's call, but not as vocal. **SIMILAR SPECIES** See Common Greenshank. **HABITAT, HABITS AND STATUS** Uncommon spring passage migrant, scarce in autumn and winter, to Mai Po NR and Deep Bay intertidal mudflats. (M)

Green Sandpiper ◾ *Tringa ochropus* 白腰草鷸 21–24cm

DESCRIPTION Medium-sized dark *Tringa*. In winter plumage upperparts and upper breast dark brown with greenish tone, and mantle lightly spotted. White supercilium restricted to front of eye, and narrow white eye-ring. Breeding plumage similar, but with all features more contrasting. In flight conspicuously bright white rump and pale lower breast and belly contrast with blackish underwing and dark upperparts to produce markedly black and white appearance. Voice: flight call a loud, melodic *tlu-twit-whit-whit*. **SIMILAR SPECIES** See Wood Sandpiper (opposite). **HABITAT, HABITS AND STATUS** Bobs when foraging. Escape flight snipe-like towering. Typically solitary. Common passage migrant and winter visitor mainly to fish ponds around Deep Bay, but also freshwater habitat elsewhere in New Territories. (M, L)

▪ SANDPIPERS & SNIPES ▪

Wood Sandpiper
▪ *Tringa glareola* 林鷸 21–24cm

DESCRIPTION Superficially similar to Green Sandpiper (see opposite), but slimmer, longer legged and paler, with more obvious spotting on upperparts and conspicuous pale supercilium. Underwing pale with little contrast to pale underparts. Legs typically green, sometimes yellowish. Voice: flight call an urgent *chiff-iff-iff*. **HABITAT, HABITS AND STATUS** Like Green Sandpiper has towering escape flight. Common passage migrant and winter visitor to freshwater habitat in New Territories. (M, L)

Grey-tailed Tattler ▪ *Tringa brevipes* 灰尾漂鷸 23–27cm

DESCRIPTION Plain grey wader with rather short legs for a *Tringa*, and longish wings and tail. Upperparts dark grey; underparts white, the breast washed grey. White supercilium and dark lores. Rather long, thickish bill, dark with a yellowish base; yellow legs. In breeding plumage neck, breast and flanks narrowly barred grey. **HABITAT, HABITS AND STATUS** Fairly common spring passage migrant, less numerous in autumn, to widespread coastal localities throughout Hong Kong, especially rocky shores. Also with concentrations in Deep Bay. (M, I)

▪ SANDPIPERS & SNIPES ▪

Terek Sandpiper ▪ *Xenus cinereus* 翹嘴鷸 22–24cm

DESCRIPTION Reminiscent of Common Sandpiper (see below), but with short orangey legs and uniquely long, upturned bill. In winter plumage has pale brown-grey upperparts and white underparts. Black carpal patch on folded wing. Bill dark with orangey base. Breeding plumage similar, with distinct black line along scapulars. In flight bold white tips to secondaries recall Common Redshank (see p. 52). **HABITAT, HABITS AND STATUS** When feeding typically employs sudden, darting runs. Common spring passage migrant, less numerous in autumn, mainly to intertidal mudflats of Deep Bay. (M)

Common Sandpiper ▪ *Actitis hypoleucos* 磯鷸 19–21cm

DESCRIPTION Small, rather long-tailed, short-legged wader. Upperparts olive-brown, pale supercilium and distinct, narrow white eye-ring. Patch on side of breast greyish-brown; rest of underparts, including obvious wedge bordering breast-patch, white. In flight shows conspicuous white wing-bar and sides of tail. Bill short and dark, with paler base. Leg colour variable, ranging from grey through greenish to yellow-brown. Voice: vocal in flight, a high-pitched *slee-lee-lee*. **SIMILAR SPECIES** Green Sandpiper (see p. 54) is darker, with blackish underwings, and white supercilium does not extend behind eye. **HABITAT, HABITS AND STATUS** Solitary. Flight distinctive, with shallow, fluttering wingbeats interspersed with glides, typically low over water. Active, persistently bobbing rear end. Common passage migrant and winter visitor with occasional summer records, mainly on coast and wetlands of north-west New Territories, but also in similar habitat throughout Hong Kong. (M, L, T, I)

▪ Sandpipers & Snipes ▪

Ruddy Turnstone ▪ *Arenaria interpres* 翻石鷸 22–24cm

DESCRIPTION Thickset, medium-sized, short-legged wader with distinctive bold, tortoiseshell pattern in breeding plumage, which many spring migrants attain in Hong Kong. Black and white-patterned head and breast, bright chestnut upperparts and white underparts. Stubby black bill with upturned tip to lower mandible; bright orange-red legs. In winter plumage is duller without any chestnut. Head and upperparts blackish-brown with whitish chin. In flight intricate black and white pattern is unique among waders of the region. **HABITAT, HABITS AND STATUS** Unusual feeding technique involves flicking over debris and stones with bill to expose live food items. Comparatively tame. Common spring passage migrant; scarce in autumn, with occasional winter records mainly on intertidal mudflats of Deep Bay and at Mai Po NR. (M)

Great Knot
▪ *Calidris tenuirostris* 大濱鷸 26–28cm

DESCRIPTION Largest calidrid. Winter plumage similar to Red Knot's (see p. 58), with longer bill, stronger patterning above and more obvious spotting on breast extending to breast sides. Long wings reach beyond tail, giving more elongated shape. In breeding plumage bold spotting on breast forms blackish breast-band merging to chevrons on flanks. Mantle feathers black with white fringes; scapulars bright chestnut edged with white. **HABITAT, HABITS AND STATUS**. Common spring migrant; less numerous in autumn with a few individuals wintering mainly on Deep Bay intertidal mudflats and at Mai Po NR. (M)

■ SANDPIPERS & SNIPES ■

Red Knot ■ *Calidris canutus* 紅腹濱鷸 23–25cm

DESCRIPTION Fairly large, stocky, short-legged calidrid. In winter essentially nondescript grey upperparts and breast; rest of underparts white. Shortish black bill; greyish-green legs. Transformed in breeding plumage, with bright chestnut face and underparts, and chequered black, white and chestnut upperparts. **HABITAT, HABITS AND STATUS** Common spring passage migrant; far less numerous in autumn and winter, almost exclusively on Deep Bay intertidal mudflats and at Mai Po NR. (M)

Sanderling ■ *Calidris alba* 三趾濱鷸 20–21cm

DESCRIPTION Medium-sized calidrid. In winter plumage very pale with silvery-grey upperparts, conspicuous blackish shoulder-patch, white underparts, short black bill and black legs. In breeding plumage upperparts, head and breast mottled chestnut, and black with white feather tips on mantle. **HABITAT, HABITS AND STATUS** Typically very active, running after waves on seashore. In Hong Kong often associates with mixed Dunlin and Curlew Sandpiper (see pp. 62 and 61) flocks on Deep Bay intertidal mudflats and at Mai Po NR. Uncommon spring passage migrant, even scarcer in autumn. (M)

Non-breeding  *Breeding*

= Sandpipers & Snipes =

Red-necked Stint ■ *Calidris ruficollis* 紅頸濱鷸 13–16cm

DESCRIPTION Along with other stints, one of the smallest waders of the region. In winter head, neck, upperparts and breast sides are pale grey. Supercilium, throat and underparts are white. In breeding plumage well-marked birds display orange-red chin, throat, ear-coverts, sides of neck and upper breast, although others show less extensive and pale red plumage. Underparts white with dark streaking on breast. Mantle and scapulars boldly patterned rufous and black, tipped white. Bill and legs black. **SIMILAR SPECIES** Both Long-toed and Temminck's Stints (see p. 60) have pale legs. See also Little Stint (below). **HABITAT, HABITS AND STATUS** One of the most numerous waders on spring passage, forming large feeding and roosting flocks. Scarce autumn passage migrant and winter visitor. Most records are from intertidal mudflats of Deep Bay and Mai Po NR. (M)

Little Stint

■ *Calidris minuta* 小濱鷸 12–14cm

DESCRIPTION Separation from Red-necked Stint (see above) requires careful observation. Subtly shorter winged and longer legged. In breeding plumage orange-red colour is less intense and restricted to ear-coverts, neck and sides of breast; coverts and primaries, as well as mantle, edged with rufous. Note white throat. In non-breeding plumage extremely similar to Red-necked Stint. Immature shows split pale supercilium and white V on mantle. **HABITAT, HABITS AND STATUS** Rare other than on spring passage, when it is an uncommon visitor to Deep Bay area. (M)

■ Sandpipers & Snipes ■

Temminck's Stint ■ *Calidris temminckii* 青腳濱鷸 13–15cm

DESCRIPTION Small. Resembles miniature Common Sandpiper (see p. 56). Plain grey-brown head and upperparts, and dingy grey-brown wash to breast. Remainder of underparts white. In flight white outer-tail feathers are distinctive. Breeding plumage somewhat brighter, with chestnut-edged scapulars. Legs usually dull brownish-yellow, sometimes tending towards greenish. **HABITAT, HABITS AND STATUS** Shows distinct preference for margins of freshwater pools in wet meadows and drained fish ponds. Tendency to tower if flushed. Common winter visitor and passage migrant. Recorded almost exclusively in northern New Territories. (M)

Long-toed Stint ■ *Calidris subminuta* 長趾濱鷸 13–14cm

DESCRIPTION Small. Marginally smaller than Red-necked Stint (see p. 59), and slimmer with longer neck and legs. In winter plumage upperparts are greyish-brown with darker feather centres. Breast brushed grey-brown with light streaking; remainder of underparts white. Bill dark with yellowish base to lower mandible. Legs yellow- to greenish-brown. In breeding plumage crown and upperparts edged with bright rufous, recalling much larger Sharp-tailed Sandpiper (see opposite). **HABITAT, HABITS AND STATUS** Habitat preference similar to that of Temminck's Stint (see above) – that is, it is less likely to be found on intertidal mudflats. Regularly adopts upright stance with neck stretched upwards. Common passage migrant and scarce winter visitor, mainly recorded in northern New Territories. (M, L)

▪ SANDPIPERS & SNIPES ▪

Sharp-tailed Sandpiper ▪ Calidris acuminata 尖尾濱鷸 17–20cm

DESCRIPTION Medium-sized wader with noticeably clear-cut dark rusty cap and pale supercilium stronger behind eye. Upperparts rich brown with buff edges; breast streaking profuse but light and merges into white belly. Bill mainly black with hint of downward curve. Grey-green legs, sometimes with yellowish tone. Crown chestnut in breeding plumage, and mantle, scapulars and tertials edged with chestnut; blackish-brown streaking on underparts forms distinct chevrons, especially on flanks. Juveniles in autumn have orange-buff wash to breast. **HABITAT, HABITS AND STATUS** Common spring passage migrant, but far less numerous in autumn, to Deep Bay area. (M)

Curlew Sandpiper ▪ Calidris ferruginea 彎嘴濱鷸 18–19cm

DESCRIPTION In winter plumage recalls Dunlin (see p. 62). More elegant, with larger yet slimmer build, longer legs and longer, finer, decurved bill. Upperparts grey; head with conspicuous broad white supercilium. White rump distinctive in flight. In breeding plumage head, mantle and entire underparts deep brick-red, relieved by white-edged grey wing-coverts. **HABITAT, HABITS AND STATUS** Passage migrant, abundant in spring when flocks in excess of 5,000 birds occur; much less numerous in autumn, mostly on intertidal mudflats of Deep Bay and at Mai Po NR (M)

▪ Sandpipers & Snipes ▪

Dunlin ▪ *Calidris alpina* 黑腹濱鷸 16–20cm

DESCRIPTION Smallish, compact wader with longish, slightly decurved black bill, black legs, rather plain greyish-brown upperparts, inconspicuous pale supercilium, breast lightly washed and streaked grey, and rest of underparts white. In breeding plumage (which some Hong Kong birds assume by April), mantle, scapulars and tertials show bright chestnut fringes, and belly has conspicuous large black patch. **HABITAT, HABITS AND STATUS** Abundant winter visitor to intertidal mudflats of Deep Bay, where it forms substantial flocks. (M)

Non-breeding  *Breeding*

Broad-billed Sandpiper ▪ *Limicola falcinellus* 闊嘴鷸 16–17 cm

DESCRIPTION Smallish, short-legged wader characterized by longish black bill with an exaggerated downward kink towards tip, and a double supercilium giving a distinctive head pattern. In breeding plumage upperparts brown with a hint of rufous, and white-edged scapulars and mantle feathers produce rather conspicuous pale lines. Breast streaked brown and distinctly defined from white belly. Winter plumage more grey-brown, with upper line of split supercilium less obvious. **HABITAT, HABITS AND STATUS** Passage migrant, common in spring, less numerous in autumn and occasional in winter, almost exclusively to intertidal mudflats of Deep Bay and Mai Po NR. (M)

■ Sandpipers, Snipes & Pratincole ■

Red-necked Phalarope ■ *Phalaropus lobatus* 紅頸瓣蹼鷸 18–19cm

DESCRIPTION Dainty, slim, rather long-necked wader with longish, fine, all-dark bill. In winter plumage crown, hindneck and upperparts grey with whitish fringes to upperparts, bold black eye-patch extending back from just in front of eye, and white underparts. Legs grey-black; feet with lobed toes. Bright breeding plumage dominated by wide orange-red stripe running from rear of face down side of neck to side of upper breast. Rest of head slate-grey with white eye-ring and throat. Upperparts rich brown, patterned with bold golden-buff line on mantle and chestnut edges to scapulars. Male performs all brooding duties and is a pale version of female. **HABITAT, HABITS AND STATUS** Spends most of its time swimming, when it is very buoyant and erratic in direction, regularly performing rapid spins and feeding by picking at water's surface. Typically extremely confiding. Common passage migrant in numbers that vary considerably from year to year. Most records are at sea, with regular reports also from fish ponds in New Territories and Hong Kong Harbour. (M, L, I)

Oriental Pratincole ■ *Glareola maldivarum* 普通燕鴴 23–24cm

DESCRIPTION Remarkably different wader with tern-like features of long, narrow wings, forked tail and short legs. Plumage largely greyish-brown, with warm orangey tones on lower breast, white belly and vent, and black primaries and tail. Handsome facial appearance based upon creamy throat bordered by narrow black band, and black bill brightened at base with lipstick-red, which extends onto gape. Conspicuous white rump and uppertail in flight, and deep tawny-red axillaries and underwing-coverts shine bright when caught in the light, yet otherwise often appear black. Plumage duller and less distinct in winter. Sexes are similar. Immature greyer than adults, with pale fringed upperparts and breast feathers. **HABITAT, HABITS AND STATUS** Gregarious. Dashing, agile flight in pursuit of insect food, with feeding flocks forming over insect swarms. Occasionally flocks seen spiralling at considerable height. Common spring passage migrant, uncommon in autumn, recorded mainly in northern New Territories. (M, L, I)

▪ Gulls & Terns ▪

Black-headed Gull ▪ *Chroicocephalus ridibundus* 紅嘴鷗 34–37cm

DESCRIPTION Smallish, pale-backed gull with white head irregularly marked blackish, but with consistent dark spot on ear-coverts and white underparts. In breeding plumage acquires dark brown hood, which can begin to appear as early as December. Bill dark red with dark tip in winter; legs red. In flight shows distinct white panel to leading edge of outer wing. First-winter birds have gingery brown-centred wing-coverts and tertials, black tail-band, and paler bill and legs than adults. **HABITAT, HABITS AND STATUS** Abundant winter visitor to coastal waters. (M)

Saunders's Gull ▪ *Chroicocephalus saundersi* 黑嘴鷗 29–32cm

DESCRIPTION Smallest and daintiest of Hong Kong's regular gulls. Pearl grey mantle and wings, black and white-tipped primaries, and white head and underparts, the former with small dark 'ear-spot' in winter. Underwing white with black rectangle on outer primaries – most underwing primaries dark in Black-headed Gull (see above) except white outers. Stubby black bill and shortish black legs. Acquires breeding plumage with black hood and conspicuous white eye-crescents early in spring, before most Black-headed Gulls. First-winter birds have similar head pattern to Black-headed Gulls of same age, and brown-centred coverts and tertials. **HABITAT, HABITS AND STATUS** Flight reminiscent of Gull-billed Tern's (see p. 66). Forages from air over mudflats, 'falling' onto prey (often crabs) in vertical stoop. Common winter visitor, exclusively to Deep Bay and Mai Po NR. (M)

▪ Gulls & Terns ▪

Black-tailed Gull
■ *Larus crassirostris* 黑尾鷗 44–47cm

DESCRIPTION Medium-sized, dark grey-backed gull with white head (streaked dark in winter) and underparts. Bill yellow with narrow black band bordering red tip. Legs yellow. In all plumages shows wide black tail-band, especially apparent in flight. In brown immatures, more frequently seen in Hong Kong than adults, tail-band is even broader and set off by white rump. Bill pink with black tip. **HABITAT, HABITS AND STATUS** Common winter visitor to Deep Bay, and widespread spring passage migrant in offshore waters. (M, I)

Caspian Gull ■ *Larus cachinnans* 蒙古銀鷗 56–68cm

DESCRIPTION Large white gull with pale grey mantle and wings, and white-tipped black outer primaries with prominent white 'mirrors' on wing-tips. In winter adult appears white headed with no streaking, except for hind-collar of faint brown streaking on rear and sides of neck. Bill yellow with red gonys spot; legs pink, not yellow, on Mongolian taxon *L. c. mongolicus*, which occurs in Hong Kong. Immature (first winter) light greyish-brown, with wholly dark outer primaries, noticeably pale head, black bill and sharply defined black tail-band contrasting with white rump. **SIMILAR SPECIES** Adult winter **Vega Gull** *L. vegae*, a scarce winter visitor to Deep Bay, has a strongly brown-streaked head, rear and sides of neck, while immature has a pale base to bill and more heavily patterned upperparts. Immature (first-winter) Heuglin's Gull (see p. 66) is similar, but has a darker upperwing in flight and broader, less well-defined, dark subterminal band to tail. **HABITAT, HABITS AND STATUS** Uncommon winter visitor to coastal waters, particularly Deep Bay. The taxonomy of large gulls is complex and subject to change. A recent alternative treatment combines **Mongolian Herring Gull** and Vega Gull under **Arctic Gull** as *L. smithsonianus mongolicus* and *L. s. vegae* respectively. (M)

Second winter

▪ GULLS & TERNS ▪

Heuglin's Gull ▪ *Larus fuscus heuglini* 烏灰銀鷗 53–70cm

DESCRIPTION Hong Kong's most common large gull. Mantle and wings dark grey with small white mirrors on wing-tips. Head white, streaked light brown in winter, especially around nape; underparts white. Bill yellow with red gonys spot; yellow legs become paler in winter. Immature has mottled brown upperparts and paler underparts, and head with a dark (first-winter) or dark-tipped (second-winter) bill; acquires grey mantle in second winter and most adult features by third winter. Leg colour usually changes from pink to yellow by second summer. **HABITAT, HABITS AND STATUS** Common winter visitor mainly to Deep Bay, and spring passage migrant. Note Heuglin's Gull is name given to Arctic taxon *L. f. heuglini* to differentiate it from **Lesser Black-backed Gull** *L. f. graellsii* of northern Europe. (M, I)

Gull-billed Tern ▪ *Gelochelidon nilotica* 鷗嘴噪鷗 35–38cm

DESCRIPTION Largish tern of sturdy appearance. Black cap, pale grey upperparts and white underparts. Tail-tip has distinct fork but no streamers. Bill heavy and black; legs black. In winter and immature plumages head is white, the crown and nape often showing dark grey streaks with variable blackish patch behind eye. **HABITAT, HABITS AND STATUS** Flies with leisurely wingbeats, and feeds by dipping like a marsh tern rather than plunge diving. Passage migrant common in spring, scarce in autumn, mainly in Deep Bay area. (M)

▪ Gulls & Terns ▪

Caspian Tern ▪ *Hydroprogne caspia* 紅嘴巨鷗 47–54cm

DESCRIPTION Huge tern approaching Heuglin's Gull (see opposite) in size. Tail with shallow fork. In breeding plumage black cap extends into shaggy nuchal crest. Head below eye level, hindneck and underparts white, and mantle and upperwings grey with black underside to outer primaries. Very heavy deep red bill, with extreme tip yellowish-white often bordered by greyish-black band. In winter and immature plumages black cap is reduced to streaking on rear crown, and a broad black mask from rear of eye meeting behind crown. Bill acquires blackish tip. **HABITAT, HABITS AND STATUS** Gull-like in flight with deliberate, unhurried wingbeats. Feeds by plunge diving. Common in spring, scarce in autumn and winter, to Deep Bay area. (M)

Greater Crested Tern ▪ *Thalasseus bergii* 大鳳頭燕鷗 46–49cm

DESCRIPTION Large tern with rather dark grey mantle and wings. In breeding plumage crown black with bushy crest at rear, and white band across forehead and lores. Bill yellow, occasionally with orange tone. In winter forecrown white, rear crown and mask from eye to nape black mottled white, and bill paler yellow, often with greenish hue. Juvenile similar to winter adult except for strongly patterned upperwing with alternating dark and light bands due to blackish forewing, greater coverts bar, primaries and secondaries. **HABITAT, HABITS AND STATUS** Common spring passage migrant in coastal waters, with occasional records in autumn and winter. Like most terns, regularly rests on flotsam. (I)

▪ Gulls & Terns ▪

Little Tern
▪ *Sternula albifrons* 白額燕鷗 22–24cm

DESCRIPTION Very small white tern with forked tail and short streamers. White forehead and line over eye. Narrow line from bill to eye, crown to eye-level and nape black. Upperwing and mantle grey; rump, tail and underparts white. Bill yellow with black tip; legs orangey-yellow. In winter plumage white of forehead extends to mid-crown with dark streaking on rear crown. Black line from rear of eye joins black nape. Rump and tail grey. Bill black and legs dirty yellow. **HABITAT, HABITS AND STATUS** Hovers and plunge dives. Rapid wingbeats. Passage migrant in coastal waters, uncommon in spring, occasional in autumn. (M, I)

Aleutian Tern ▪ ▪ *Onychoprion aleuticus* 白腰燕鷗 32–34cm

DESCRIPTION Similar in size to Common Tern (see p. 70), but with slightly longer bill and wings. Forehead and line extending to above eye white, sharply demarcated from black lores, eye-stripe and crown. Chin, upper breast, collar, rump, vent and tail white; rest of body grey. Underwing white with blackish outer primaries and bar across secondaries. Bill and legs black. Winter adult has extensive white forehead and crown merging into dark nape, similar to other winter terns. Juvenile has dark brown back with cinnamon fringes. **SIMILAR SPECIES** Adult Common Terns in Hong Kong of East Asian taxon *longipennis* have grey underparts, black legs and similar but less contrasting underwings. Aleutian Tern best separated in spring and early autumn by clear-cut white forehead and line above eye. **HABITAT, HABITS AND STATUS** Uncommon passage migrant in southern waters. Numbers vary from year to year, with spring passage heavier than autumn one. (I)

Gulls & Terns

Bridled Tern ■ *Onychoprion anaethetus* 褐翅燕鷗 30–32cm

DESCRIPTION Heavier and longer winged than Black-naped Tern (see p. 70), with slower and more purposeful flight. From above, the darkest tern occurring regularly in Hong Kong. Upperparts including wings, back, rump and tail blackish-grey, appearing dark brown in strong light. Crown, nape and line from bill back through eye black, with white line across forehead extending back to above eye. Underparts white. Legs and bill black. Winter adult paler, usually with whitish edges to mantle and scapulars. Juvenile similar but with less well-defined head markings and broader pale edges to upperparts. **HABITAT, HABITS AND STATUS** Oceanic. Feeds by dipping in the manner of Gull-billed Tern (see p. 66), but also dives from low height. Common summer breeder on islands in Mirs Bay, and passage migrant. High numbers have occurred after typhoons. (I)

Roseate Tern ■ *Sterna dougallii* 粉紅燕鷗 35–38cm

DESCRIPTION Elegant pale sea tern similar to Common Tern (see p. 70), but with shorter and paler wings, longer bill and legs, much longer tail streamers, and paler mantle. Upperparts pearl-grey, appearing whitish in strong light, with little contrast against white rump and tail. Underparts white; breast sometimes with pink suffusion. Crown, nape and hindneck black. Slender bill orange-red with black tip. Legs red. Winter adult has white lores, forehead and forecrown, and dark bar across lesser coverts. Juvenile has bold dark scaling on grey upperparts. **SIMILAR SPECIES** See Black-naped (see p. 70) and Common Terns. **HABITAT, HABITS AND STATUS** Feeds by plunge diving. Summer breeding visitor to islands in eastern and southern waters, in smaller numbers than Black-naped and Bridled Terns (see above). (I)

▪ GULLS & TERNS ▪

Black-naped Tern ▪ *Sterna sumatrana* 黑枕燕鷗 34–35cm

DESCRIPTION Slender sea tern with very pale appearance and buoyant flight. Head and underparts white relieved by narrow black band from eye circling nape and extending down hindneck. Upperparts very pale grey; rump and tail white with long streamers. Bill and legs black. Winter plumage similar with some dark streaking on crown. Juvenile shows less defined black band and nape, buffish head, brownish-black chevrons on mantle and pale grey underparts. **HABITAT, HABITS AND STATUS** Plunge dives. Common summer visitor and passage migrant, breeding on rocky islets in eastern and southern waters. (I)

Common Tern ▪ *Sterna hirundo* 普通燕鷗 32–38cm

DESCRIPTION Graceful sea tern approaching size of Black-headed Gull (see p. 64), but structure is slender and attenuated, enhanced by long tail streamers. In breeding plumage cap and nape are black; head below eye level, sides of neck, underparts, rump and tail are white. Upperparts pale grey. Two, possibly three, taxa occur in Hong Kong, best distinguished by bill and leg colour, black in *S. h. longipennis* and red (bill with black tip) in *S. h. tibetana* and *minussensis*; separation of the last two is not possible in the field. In winter and juvenile plumages forehead and lores are white and lesser coverts blackish, producing a dark bar at bend of wing. Outer primaries form dark wedge. Bill black, usually with reddish base in *tibetana/minussensis*. **HABITAT, HABITS AND STATUS** Feeds by plunge diving. Uncommon passage migrant in coastal waters. (M, I)

GULLS & TERNS

Whiskered Tern ■ *Chlidonias hybrida* 鬚浮鷗 23–25cm

DESCRIPTION Smallish tern, slightly larger than White-winged Tern (see below). In breeding plumage almost entire body is smoky-grey, and even darker on belly, enhancing contrast with white cheeks and ventral area. Underwings pale grey. Well-defined black cap and white cheeks. Tail slightly forked, lacking streamers. Bill and legs dark red. In winter and immature plumages, upperparts pale grey, head and underparts white, and rear crown streaked black merging into dark patch behind eye. Bill and legs blackish, although adults can show reddish suffusion. **SIMILAR SPECIES** See White-winged Tern. **HABITAT, HABITS AND STATUS** One of the so-called marsh terns, feeding on aerial insects and by dipping to surface of water. Common passage migrant mainly to fish ponds and intertidal zone of Deep Bay area, including Mai Po NR and coastal waters. (M, I).

White-winged Tern ■ *Chlidonias leucopterus* 白翅浮鷗 20–23cm

DESCRIPTION Smallest marsh tern. In breeding plumage strikingly black and white in flight. White upperwing-coverts, vent, rump and tail contrast with otherwise black plumage, except for grey inner primaries and outer secondaries. Underwing pattern reversed, with black coverts and light grey flight feathers. Bill dark blackish-red; legs red. Winter plumage similar to that of Whiskered Tern (see above), and best distinguished by stubbier, black bill, pale rump and more extensive black central crown-patch that reaches behind and below eye level. **HABITAT, HABITS AND STATUS** Passage migrant common in spring, less numerous in autumn, to Deep Bay area and fish ponds in northern New Territories as well as at sea. (M, I)

◾ Jaeger & Auk ◾

Long-tailed Jaeger ◾ *Stercorarius longicaudus* 長尾賊鷗 48–53cm

DESCRIPTION Vaguely gull-like seabird close in size to Black-headed Gull (see p. 64), but with unusually long, pointed tail streamers that account for more than one-third of its length. In breeding plumage black cap extends below eye to rear neck; lower 'face', throat and neck collar yellowish-cream fading to white on breast. Remainder of body greyish-brown; flight feathers black. In winter plumage cap indistinct, and head and throat browner; upperparts barred and lacks tail streamers. Immature duller that adult and more extensively barred. **HABITAT, HABITS AND STATUS** Oceanic. Flight agile and graceful. Chases terns to force them to drop or disgorge fish. Uncommon offshore passage migrant, mainly in spring. (I)

Ancient Murrelet ◾ *Synthliboramphus antiquus* 扁嘴海雀 24–27cm

DESCRIPTION Small, plump, bull-necked auk. In breeding plumage black head and throat are relieved by white eyebrow plumes extending to nape. Upperparts slaty-grey; underparts white, including connected patch at side of neck. Stubby bill ivory coloured. In winter lacks head plumes and throat is less extensively black. **HABITAT, HABITS AND STATUS** Oceanic. Swims and dives frequently when feeding. Fast, whirring flight. Uncommon offshore spring passage migrant, rarely in winter. (I)

PIGEONS & DOVES

Domestic Pigeon
■ *Columba livia* 原鴿 31–34cm

DESCRIPTION Feral descendant of wild Rock Dove of Eurasia, with random plumage patterns largely restricted to black, grey, brown and white of all tones. Some individuals have iridescent green, blue and/or purple suffusion on head, neck and/or upper body. Classic Rock Dove plumage combines grey head and body with white band across lower back and upper rump, and conspicuous black double wing-bars. Neck, upper mantle and upper breast glossed green or purple depending on angle of light. **HABITAT, HABITS AND STATUS** Common resident in widespread urban areas. (M, L, H, I)

Oriental Turtle Dove ■ *Streptopelia orientalis* 山斑鳩 33–35cm

DESCRIPTION Grey crown and hindneck with obvious black and white neck-bars. Face and underparts pale vinaceous-pink, lighter on belly and ventral area. Mantle, scapulars and wing-coverts deep rufous with black feather centres. Rump blue-grey. Flight feathers and tail blackish-grey, the tail with a narrow whitish-grey band at tip. **HABITAT, HABITS AND STATUS** Common winter visitor occurring throughout Hong Kong wherever there is tree cover. Almost certainly breeds in some years. (M, L, I)

PIGEONS & DOVES

Eurasian Collared Dove ■ *Streptopelia decaocto* 灰斑鳩 31–33cm

DESCRIPTION Plain, pale greyish-brown dove with white-edged black half-collar on hindneck. Undertail shows broad white terminal band to black tail. Juvenile lacks half-collar. **HABITAT, HABITS AND STATUS** Recent colonist; now common resident in north-west New Territories. (M)

Red Turtle Dove ■ *Streptopelia tranquebarica* 火斑鳩 20.5–23cm

DESCRIPTION Distinctly smaller than other *Streptopelia* doves in Hong Kong. Sexes separable. Male has pale grey head and pinkish-brown body, deeper on upperparts and wings. Narrow black half-collar encircles hindneck. Rump and tail grey. Outer-tail feathers tipped white. Female similar but less colourful, with brown rather than pink tones. **HABITAT, HABITS AND STATUS** Common passage migrant and winter visitor. Widespread throughout Hong Kong, although favours Deep Bay area. (M, L, I)

▪ Pigeons & Doves ▪

Spotted Dove
▪ *Spilopelia chinensis* 珠頸斑鳩 27.5–30cm

DESCRIPTION The common dove in Hong Kong. Grey crown and face; grey-brown upperparts, mantle feathers, scapulars and wing-coverts broadly edged with pale vinaceous-brown. Throat, nape and underparts deep vinaceous-pink. Rather long, blackish-brown tail shows white tips to outer feathers. Wide black half-collar copiously spotted white. Juvenile plain greyish-brown with pale forehead. Lacks half-collar. **HABITAT, HABITS AND STATUS** Abundant and widespread resident, including in urban areas. (M, L, T, H, I)

Common Emerald Dove ▪ *Chalcophaps indica* 綠翅金鳩 23–27cm

Seen in MunyTAK

DESCRIPTION Secretive woodland dove. Male has white forehead and eyebrows, lavender crown and hindneck, bright green upperparts, and deep vinaceous lower 'face', neck sides and breast merging to pinkish-brown belly. Carpal-bar white with grey upper border. In flight two white bars conspicuous on grey lower back. Bill and legs bright red. Female duller than male and lacks carpal-bar. Juvenile darker and browner than adults; forehead and crown brown, underparts barred rufous-brown and green plumage restricted to upperwing-coverts. **HABITAT, HABITS AND STATUS** Uncommon but widespread resident in closed woodland. (T, I)

■ Cuckoos ■

Greater Coucal ■ *Centropus sinensis* 褐翅鴉鵑 47–52 cm

DESCRIPTION Large, heavily built ground cuckoo with long, broad tail. Glossy black with chestnut mantle and wings. Stout black bill, red iris and black legs. Juvenile's black plumage barred white, and brown wings and mantle closely barred black. Voice: one of the most distinctive natural sounds of Hong Kong. Far-carrying, hollow *whoop, whoop, whoop, whoop…* in descending series of about ten accelerating notes. **HABITAT, HABITS AND STATUS** Common resident. Widespread throughout Hong Kong in scrub, grassy hillsides and mangroves. Often walks on the ground. Flight heavy and laboured and typically only for short distance. Usually remains hidden within bush. (M, L, I)

Lesser Coucal ■ *Centropus bengalensis* 小鴉鵑 33cm

DESCRIPTION Small edition of Greater Coucal (see above) with chestnut underwing-coverts (black in Greater) and less heavy bill. Juvenile has pale underparts and blackish barred rufous-brown tail. Voice: similar to Greater Coucal's, but higher pitched and noticeably less resonant, followed by harsh, repeated *kokoruk*. **HABITAT, HABITS AND STATUS** Uncommon and widespread resident. Favours grassy and scrubby hillsides, often at higher altitude than Greater Coucal, and overgrown open areas. (M, L, I)

▪ Cuckoos ▪

Chestnut-winged Cuckoo
▪ *Clamator coromandus* 紅翅鳳頭鵑 46 cm

DESCRIPTION Striking large, very long-tailed cuckoo. Head, including untidy, upstanding crest, mantle, lower belly, vent and tail black; orange-buff chin and throat, and white half-collar, breast and upper belly. Wings bright rufous-brown. Voice: loud, monotonous two-note call, *beep-beep*, repeated for long periods. Also chatters.
HABITAT, HABITS AND STATUS Uncommon and widespread summer visitor and passage migrant, scarcer in autumn. Favours woodland and thick, shrubby areas. Brood parasite. (M, T, I)

Asian Koel ▪ *Eudynamys scolopaceus* 噪鵑 43cm

DESCRIPTION Large, long-tailed cuckoo. Male entirely glossy black, relieved only by bright red iris and light green bill. Female has dark brown upperparts liberally spotted (barred on tail) white. Underparts white, throat spotted, and breast, belly and vent barred brownish-black. Voice: very vocal; call a repeated loud *ko-el*, rising in pitch. Also a slightly bubbly, rising *clu, clu, lu, lu, lu, lu*, varying in number of notes. **HABITAT, HABITS AND STATUS** Common in summer, scarce in winter. Widespread throughout Hong Kong wherever there are trees, including in urban areas. Brood parasite. (M, L, T, H, I)

Male

Female

■ Cuckoos ■

Plaintive Cuckoo
■ *Cacomantis merulinus* 八聲杜鵑 18–23.5cm

DESCRIPTION Small cuckoo with grey head, neck, mantle and upper breast, greyish-brown upperparts, and orange-buff lower breast, belly and vent. Dark undertail with broad white tips and barring. Juvenile grey-brown above with orange-buff barring on wings and tail, and brown, lightly barred white underparts. First-winter birds similar to juveniles, but underparts wholly orange-buff. Voice: song variable, but version most regularly heard in Hong Kong is a loud, high-pitched, descending, whistled *heer, heer, heer, heer, heer-a-he-a-he*. **HABITAT, HABITS AND STATUS** Common spring and summer visitor, less frequently recorded in autumn and winter, to widespread open areas. Brood parasite. (L, I)

Large Hawk Cuckoo ■ *Hierococcyx sparverioides* 大鷹鵑 38–40cm

DESCRIPTION Largish cuckoo with grey head, greyish-brown upperparts, and long, grey and black-banded tail. Underparts white, throat streaked, and breast and belly barred dark grey. Upper breast and, variably, throat orange-buff. Voice: endlessly repeated, loud, whistled, three-note call, *whee, whee-ar*, or *brain fever*. **SIMILAR SPECIES** See Hodgson's Hawk Cuckoo (opposite). **HABITAT, HABITS AND STATUS** Common spring and summer visitor to widespread closed woodland. Can be frustratingly difficult to see even when calling at close range. Brood parasite. (L, T, I)

▪ Cuckoos ▪

Hodgson's Hawk Cuckoo
▪ *Hierococcyx nisicolor* 霍氏鷹鵑 28–30cm

DESCRIPTION Considerably smaller than Large Hawk Cuckoo (see opposite). Head and upperparts slate-grey, with variable white spot on nape. Tertials pale pearly grey. Long tail banded slate and pale grey, with narrow white tip bordered above with rufous. Chin slate; throat white. Underparts white, breast copiously streaked orange-brown. Yellow eye-ring; bill-base yellow, tip black. Immature has bold blackish streaking, edged with rufous, on underparts. Wings and mantle dark brown edged with orange-brown. Voice: repeated two-note *gee-wiz*, rising in volume and pitch. **HABITAT, HABITS AND STATUS** Uncommon spring and summer visitor to enclosed woodland and extensive shrubbery in New Territories. Brood parasite. (T, I)

Indian Cuckoo ▪ *Cuculus micropterus* 四聲杜鵑 32–33cm

DESCRIPTION Medium-sized cuckoo. Pale grey head, neck and upper breast contrasting with dark brown upperparts and tail, the latter with wide black terminal band. Underparts white, conspicuously barred black. Iris brown. Female has rufous wash on throat and breast. Voice: repeated, loud, high-pitched *uoo-uoo-uoo-oo*, the last note lower than the first three. **HABITAT, HABITS AND STATUS** Common spring and summer visitor. Widespread wherever there is tree cover. Brood parasite. (M, L, I)

■ Owls ■

Collared Scops Owl ■ *Otis lettia* 領角鴞 23–25cm

DESCRIPTION Smallish owl with short 'ear-tufts', buff 'eyebrow' and narrow black border to sides of facial disc. Upperparts grey-brown, mottled darker with sparse white spotting; patches on scapulars show as a diagonal row of large spots. Pale rear half-collar. Underparts buffish-white finely vermiculated and boldly streaked black. Iris dark reddish-brown. Voice: often-repeated, rather slurred, muffled single note, *hoo*. HABITAT, HABITS AND STATUS Fairly common in wooded areas, including in urban locations. Widespread. Strictly nocturnal. (T, I)

Asian Barred Owlet ■ *Glaucidium cuculoides* 斑頭鵂鶹 22–25 cm

DESCRIPTION Smallish 'neckless' owl with distinctly rounded head barred greyish-brown and buff. No 'ear-tufts'. Upperparts darker brown barred buff; white edges of scapulars and lesser coverts form 'tramlines'. Underparts barred dark brown and buff, split by whitish vertical line. Iris yellow. Voice: continuous, rapid, tremulous *oo-oo-oo-oo-oo…*, gradually rising in volume and falling in pitch at end. HABITAT, HABITS AND STATUS Common resident in wooded areas in widespread localities, mainly in central and north New Territories. Partially diurnal. (M, T)

Owls & Nightjars

Northern Boobook
■ *Ninox japonica* 鷹鴞 27–33cm

DESCRIPTION Medium-sized, rather long-tailed owl, with rounded head and slim, upright posture. Head dark brown with small white flash above bill. Upperparts dark brown with white banding on tail; underparts off-white heavily streaked rufous-brown. Iris bright orange-yellow; bill grey. **HABITAT, HABITS AND STATUS** Uncommon passage migrant mainly in spring to widespread wooded areas. (I)

Savanna Nightjar ■ *Caprimulgus affinis* 林夜鷹 20–26cm

DESCRIPTION Upperparts mottled brown, grey and black; speckled white, pale nuchal collar; pale edges to scapulars form broken line. Underparts finely barred brown, white and yellowish-buff, with small white patch on sides of lower throat. Male has white outer-tail feathers and a conspicuous white patch on outer four primaries. On female patch is pale buff and tail lacks white outer edges. Voice: high-pitched, loud, rather staccato *chweech*. **SIMILAR SPECIES Grey Nightjar** *C. jotaka*, a scarce spring migrant and occasionally recorded in summer, is darker and greyer with an entirely different call, a fairly rapid series of loud, hollow notes, *chewk, chewk, chewk, chewk, chewk, chewk…*

HABITAT, HABITS AND STATUS Uncommon but widespread resident on grassy hillsides and open areas with low vegetation and a sprinkling of pines. Nocturnal. Normally begins calling shortly after sunset. (M, I)

SWIFTS

Pacific Swift
■ *Apus pacificus* 白腰雨燕 17–18cm

DESCRIPTION Large, dark brown swift with pale chin and throat, and white rump. Tail deeply forked. At close range underparts show pale barring. Voice: high-pitched, screeching trill. **HABITAT, HABITS AND STATUS** Aerial. High-speed, often erratic flight attained with rapid wingbeats interspersed with glides. Common spring migrant and summer visitor; irregular in autumn and winter, mostly to Deep Bay area. Breeds locally on sea cliffs and possibly mountain crags. (M, I)

House Swift ■ *Apus nipalensis* 小白腰雨燕 12cm

DESCRIPTION Small dark swift with short, square-ended tail, and white forehead, throat and wide rump-patch that spills over onto flanks. Voice: high-pitched, shrill chirruping. **HABITAT, HABITS AND STATUS** Common resident and abundant spring passage migrant, with flocks of up to 1,000 birds not unusual. Widespread urban breeding colonies, usually under overhanging buildings. (M, L, T, H, I)

ROLLER & KINGFISHERS

Oriental Dollarbird ▪ *Eurystomus orientalis* 三寶鳥 27–32cm

DESCRIPTION Large-headed, thickset, short-tailed roller, similar in size to Black-capped Kingfisher (see p. 84). Head blackish-brown; throat bright blue; body and wing-coverts deep greenish-blue; flight feathers blue with conspicuous pale blue patch on primaries. Narrow red eye-ring. Broad-based, heavy bill bright red; legs dark red. Sexes are similar. Juvenile duller and browner than adult, with blackish bill. **HABITAT, HABITS AND STATUS** Sits for long periods on high, exposed perch; catches insect prey in flight. Common passage migrant. Widespread, mainly in woodland areas. (T, I)

White-throated Kingfisher ▪ *Halcyon smyrnensis* 白胸翡翠 27cm

DESCRIPTION Largish kingfisher with bright brown head, upper mantle, lesser coverts, breast sides and lower underparts. Back, rump, tail and most of closed wing bright sky-blue. Throat, breast and central belly white. Large white patch at base of primaries conspicuous in flight. Massive bright red bill, and red legs and feet. Voice: loud, high-pitched, descending trill. **HABITAT, HABITS AND STATUS** Common resident; numbers augmented in autumn and winter. Widespread in mangroves, mudflats and fish ponds in winter, largely transferring to woodland areas in summer. (M, L, T, I)

▪ Kingfishers ▪

Black-capped Kingfisher ▪ *Halcyon pileata* 藍翡翠 30cm

DESCRIPTION Largish, bright kingfisher. Black head; white collar, throat and upper breast; remainder of underparts orange. Upperparts dark purplish-blue with large black patch on wing-coverts. Substantial bright red bill; red legs and feet. In flight shows conspicuous white patch at base of primaries. Juvenile duller than adult, with thin vermiculations to sides of throat and breast. Voice: loud, high-pitched, staccato *cha-cha-cha-cha-cha-cha-cha* of uneven rhythm. **HABITAT, HABITS AND STATUS** Uncommon passage migrant and winter visitor, infrequent in summer. Widespread, principally in mangroves, tidal mudflats and fish ponds of New Territories. (M)

Common Kingfisher ▪ *Alcedo atthis* 普通翠鳥 16–17cm

DESCRIPTION Brilliantly coloured small kingfisher. Crown and wings bright blue; mantle, rump and tail paler but even brighter blue. Orange underparts and patch behind and below eye. Long, rapier-like black bill in males; bill black with a red base in females. Legs and feet red. Voice: high-pitched piping, usually in flight. **HABITAT, HABITS AND STATUS** Common. Most numerous on passage, but also present in both summer and winter at fish ponds, marshes, mangroves, streams and tidal shorelines, mainly in Deep Bay area and New Territories coast, although widespread throughout Hong Kong. Feeds on small fish caught by plunge diving either from perch or following hover. (M, L, H, I)

■ Kingfishers & Bee-eater ■

Pied Kingfisher
■ *Ceryle rudis* 斑魚狗 26cm

DESCRIPTION Medium-sized black and white kingfisher. Crown, body, crest, wide eye-stripe and nape black. Upperparts black, with white-edged feathers creating a scalloped pattern. Underparts largely white. Male with double black breast-band, the upper band broader than the lower; female with single wide, broken band. Long, dagger-like black bill. **HABITAT, HABITS AND STATUS** Feeds predominantly on fish caught by plunge diving from sustained hover with rapid wingbeats or direct from perch. Common local resident almost exclusively in northern New Territories, at fish ponds, reservoirs and coastal bays (both fresh and salt water). (M, L)

Blue-tailed Bee-eater
■ *Merops philippinus* 栗喉蜂虎 29cm, including central rectrices projection of 7cm

DESCRIPTION Medium-sized, slender, elegant and multicoloured bee-eater. Upperparts bright green; lower back, rump and tail bright light blue. Wide eye-stripe black bordered below by narrow pale blue line. Chin yellow; throat brownish-orange; breast and belly green merging to blue vent. Longish, gently curved black bill. Long, pointed, protruding central tail feathers. **HABITAT, HABITS AND STATUS** Flight dashing in pursuit of aerial insects. Perches conspicuously in upright posture, favouring overhead wires and bare branches. Uncommon and widespread passage migrant, often in small groups. (M, L, I)

Hoopoe & Asian Barbet

Eurasian Hoopoe ■ *Upupa epops* 戴勝 26–28cm

DESCRIPTION Unmistakable and unique. Head, mantle and underparts orangey-buff; back, wings and tail strikingly barred black and white. Conspicuous black-tipped crest. Bill long and gently decurved. Voice: far-carrying, hollow *poo, poo, poo*. **HABITAT, HABITS AND STATUS** Uncommon and widespread winter visitor that can appear almost anywhere, even in urban parks. Has bred in area. Inhabits open grass and scrubby areas, where it forages on the ground. Fluttering flight often likened to a butterfly's. Crest regularly raised on alighting. Walks busily when feeding, periodically probing earth. (M, L, I)

Great Barbet ■ *Megalaima virens* 大擬啄木鳥 32–35cm

DESCRIPTION The largest Asian Barbet. Dark blue head and throat (appear black at a distance); brown breast and mantle; buffish-yellow underparts heavily streaked brown; scarlet undertail-coverts, and green back, wings and tail. Stout ivory bill with black culmen and tip. Voice: loud, high-pitched, gravelly '*keea*' with a downward inflection. Far-carrying calls repeated at constant rhythm for several minutes are characteristic sound in mature woodland. **HABITAT, HABITS AND STATUS** Secretive; more often heard than seen. Undulating flight. Uncommon resident in woodland areas in central and south-eastern New Territories, but principally Tai Po Kau. (T)

▪ Woodpecker & Falcons ▪

Eurasian Wryneck ▪ *Jynx torquilla* 蟻鴷 16–17cm

DESCRIPTION Cryptically plumaged, atypical woodpecker. Mottled grey, black and brown head and upperparts. Buffy throat and upper breast; remainder of underparts off-white, all with black scalloping. Dark bill short and pointed. Voice: loud, rapid, high-pitched series of about 12–14 notes, *kee, kee, kee, kee…* **HABITAT, HABITS AND STATUS** Perches like a passerine; does not use tail as prop and does not drum. Frequently twists head. Often forages on the ground. Uncommon winter visitor and passage migrant to open areas with scattered cover. Widespread, although mostly recorded in north-west New Territories. (M, L, T, I)

Common Kestrel
▪ *Falco tinnunculus* 紅隼 32–35cm

DESCRIPTION Medium-sized, colourful falcon. Male has grey head with pale cheeks and indistinct black moustachial stripe; upperparts bright rufous-brown with sharply defined black spots and black primaries. Rump and tail blue-grey, the latter with a wide, black terminal band and white tip. Female and immature less bright, with brown head and strongly barred upperparts and spotted underparts. **HABITAT, HABITS AND STATUS** Predisposed to lengthy hovering when foraging. Common autumn migrant and winter visitor to widespread open areas. (M, L, I)

▪ FALCONS ▪

Amur Falcon ▪ *Falco amurensis* 阿穆爾隼 28–32 cm

DESCRIPTION Medium-sized falcon; sexually dimorphic. Male grey with darker eye-patch and moustachial stripe, and paler underparts. Thighs, vent and uppertail-coverts burnt orange. Underwing shows conspicuously white coverts that contrast with blackish flight feathers. Bill grey with orange-red cere. Eye-ring and legs orange-red. Female strikingly different from male, reminiscent of Eurasian Hobby (see below). Dark grey upperparts, lightly barred black; well-defined moustachial and white cheeks; off-white underparts copiously streaked and barred black; buff thighs and vent. Underwings finely barred black and white. Immature male similar to adult female, but with deeper coloured thighs and vent, and grey-washed breast and belly. Juvenile like female, but with broad buff fringes to upperparts. **SIMILAR SPECIES** Females and immatures similar to Eurasian Hobby. Amur Falcons best separated in flight by whiter, less barred underwing-coverts. **HABITAT, HABITS AND STATUS** Sometimes hovers. Uncommon autumn migrant mainly to northern New Territories. (M, L, I)

Male    *Female*

Eurasian Hobby ▪ *Falco subbuteo* 燕隼 30–36cm

DESCRIPTION Medium-sized falcon with long wings and shortish tail. Black cap extends below eye and forms narrow but distinct moustache. White cheeks, throat and small patch on either side of nape. Upperparts dark slate; tail slightly paler. Underparts' ground colour buffish-white heavily streaked black; bright rufous thighs and undertail-coverts. Sexes and juvenile similar, although female larger and juvenile more brown toned. **SIMILAR SPECIES** See Amur Falcon (above). **HABITAT, HABITS AND STATUS** Aerial feeder on hirundines and dragonflies. Flight dashing and rapid; rather narrow, curved wings and shortish tail, together with shallow wing action, recall swifts. Uncommon autumn passage migrant, even less frequent in spring and summer, to widespread open country mainly in New Territories. (M, L, I)

▪ Falcons & Cockatoo ▪

Peregrine Falcon ▪ *Falco peregrinus* 遊隼 38–48cm

DESCRIPTION Largish, powerfully built, rather short-tailed falcon. Adult has slate-grey upperparts, blacker crown, nape and moustachial, white cheeks and throat, and off-white underparts finely barred slate. Female is larger than male. Immature has browner-toned upperparts than adult, and buff underparts streaked dark brown. **HABITAT, HABITS AND STATUS** Extremely fast flight driven by rapid, shallow wingbeats interspersed with short glides. Also feeds by stooping at high speed from height. Common and widespread, including in urban areas. Regular in Deep Bay area. Resident and winter visitor. (M, L, H, I)

Yellow-crested Cockatoo
▪ *Cacatua sulphurea* 小葵花鳳頭鸚鵡 *c.* 33cm

DESCRIPTION Unmistakable. Large, and all white except for bright yellow, erectile crest and paler yellow ear-covert patch, underwing-coverts and proximal portion of undertail. Large, powerful black bill, grey feet and pale blue bare skin narrowly surrounds eye. Voice: loud, raucous screeching. **HABITAT, HABITS AND STATUS** Flight fast and somewhat fluttery with rapid, shallow, stiff wing-beats interspersed with glides. Sociable, often in small groups. Introduced, self-sustaining population of less than 100 individuals largely restricted to northern Hong Kong Island, particularly Hong Kong Park, the Zoological and Botanical Gardens, and Happy Valley. (H)

▪ Old World Parrots ▪

Alexandrine Parakeet
▪ *Psittacula eupatria* 亞歷山大鸚鵡 50–62cm, including tail of 22–36cm

DESCRIPTION Large, bright green parakeet with very long green tail (with yellowish outer rectrices), merging to greenish-blue distally and yellow tip. Head green; nape and lower 'face' pinkish-grey. Black chin extends as narrow line to reddish-pink hind-collar. Massive bill bright red. Deep red shoulder-patch. Female smaller than male; less bright and lacks black chin and red collar. Immature similar to female. **SIMILAR SPECIES** See Rose-ringed Parakeet (below). **HABITAT, HABITS AND STATUS** Gregarious. Hong Kong birds are all regarded as escapes from captivity. Uncommon resident principally in Kowloon Park, with widespread additional records. (H)

Rose-ringed Parakeet ▪ *Psittacula krameri* 紅領綠鸚鵡 37–43 cm

DESCRIPTION Smaller than Alexandrine Parakeet (see above). Male green, with black chin extending to meet narrow pink hind-collar. Rear crown, nape and long tail bluish. Bill bright red. Female similar, but lacks chin and collar markings, although sometimes shows yellowish-green collar. **HABITAT, HABITS AND STATUS** Locally common resident deriving from introduced stock. Principally occurs in parks on Hong Kong Island, with widespread additional records. (H)

▪ Cuckooshrike & Minivets ▪

Black-winged Cuckooshrike
■ *Coracina melaschistos* 暗灰鵑鵙 19.5–24cm

DESCRIPTION Medium-sized dark bird with a longish tail. Male dark grey with black wings and blackish patch around eye. Belly paler grey, merging into even paler undertail-coverts. Tail black tipped white; feathers of undertail with broad white tips. Bill strong and black; iris ruby. Female paler than male, with lighter grey-washed underparts, faintly barred darker grey, and broken white eye-ring. **HABITAT, HABITS AND STATUS** Common passage migrant and far less numerous winter visitor to widespread wooded areas. (L, T, I)

Ashy Minivet
■ *Pericrocotus divaricatus* 灰山椒鳥 18 cm

DESCRIPTION Slim, long-tailed, grey and white bird. Male has white forecrown, sides of neck and underparts. Narrow mask extends from above bill back through eye, expanding onto rear crown and nape. Upperparts grey, flight feathers and tail black, the latter with prominent white outer rectrices. Female lacks white forecrown; black crown and nape replaced with grey. Black stripe from bill to eye faintly bordered white above. **HABITAT, HABITS AND STATUS** Uncommon passage migrant, scarcer in autumn, to widespread wooded areas. (M, I)

▪ MINIVETS ▪

Grey-chinned Minivet ▪ *Pericrocotus solaris* 灰喉山椒鳥 17–19cm

DESCRIPTION Brightly coloured, small minivet. Male has blackish crown, upperparts and tail, the last conspicuously edged scarlet. Rump scarlet; underparts and patch extending from primaries across greater wing-coverts orange-red; ear-coverts grey; chin and throat paler grey. Female paler than male, with grey head including forehead; greenish-toned grey upperparts, and white chin and throat. Bright yellow underparts, wing-patch, rump and outer rectrices. **SIMILAR SPECIES** See also Scarlet Minivet (below). **HABITAT, HABITS AND STATUS** Common resident confined to well-wooded locations in New Territories, particularly Tai Po Kau. Often forms mixed flocks with Scarlet Minivet. (T)

Female *Male*

GARDEN 13/03/20

Scarlet Minivet ▪ *Pericrocotus speciosus* 赤紅山椒鳥 17–22cm

DESCRIPTION Similar to slightly smaller Grey-chinned Minivet (see above). Male has black throat, and bright scarlet underparts, upper back, rump and outer rectrices; more extensive wing-patch and tips of tertials and inner secondaries, this last feature being diagnostic. Female has yellow forecrown, chin, throat, underparts, outer rectrices, rump and wing-patches. **HABITAT, HABITS AND STATUS** Common resident in widespread wooded areas, especially Tai Po Kau. (T)

Female  *Male*

▪ SHRIKES ▪

Brown Shrike ▪ *Lanius cristatus* 紅尾伯勞 18 cm

DESCRIPTION Forehead and crown pale grey, gradually becoming grey-brown at rear and nape. Bold black face-mask with thin whitish supercilium; white lower 'face', chin and throat. Upperparts grey-brown, rump rufous-brown, wings blackish-brown and tertials edged whitish-buff. Tail dark rufous-brown narrowly edged rufous-buff. Underparts washed buff, deeper on flanks. Stout black bill noticeably hooked. Male *L. c. lucionensis* is described here. Female similar but duller than male, sometimes with brown-tipped flank feathers producing scalloped effect. First-winter individuals like female, but with more extensive scalloping on underparts. **HABITAT, HABITS AND STATUS** Solitary. Perches prominently, dropping or swooping to the ground to catch prey. Race *lucionensis* is a common passage migrant and far less frequent winter visitor to widespread open areas. More rufous-brown *cristatus* is a scarce passage migrant mostly in autumn. Very similar *confusus* (paler upperparts and more extensive white forehead) has also been recorded. (M, L, I)

Long-tailed Shrike ▪ *Lanius schach* 棕背伯勞 20–25cm

DESCRIPTION Large shrike with noticeably long tail. Black facial mask extends broadly onto forehead. Crown, nape and upper back grey. Mantle, rump, flanks and undertail-coverts burnt orange. Throat, lower 'face' and underparts white. Wings black, tertials edged orange-buff and small white patch at base of primaries. Tail black; outer rectrices edged orange-buff. First-winter birds duller than adults; browner with distinct brown scalloping on mantle, rump, wing-coverts and underparts. Melanistic morph '*fuscatus*' has iron grey crown, nape, mantle and underparts, the last sometimes with a brownish tone. Black forehead, 'face', throat, upper breast, wings and tail. Rump and undertail-coverts sometimes orangey-brown. **HABITAT, HABITS AND STATUS** Typically hunts from exposed perch. Common resident in widespread open areas. Melanistic morph fairly frequent in northern New Territories. (M, L, I)

Melanistic morph

Typical form

VIREO & ORIOLE

White-bellied Erpornis ■ *Erpornis zantholeuca* 白腹鳳鶥 11–13cm

DESCRIPTION Small, leaf warbler-sized passerine. Pale yellowish-green above, greyish 'face' and underparts, and yellow undertail-coverts. Short, thick crest. Pale pink legs. Voice: song a high-pitched, rapid trill, *zee-zee-zee-zee-zee-zee-zee*… Calls scolding, nasal, tit-like *ti-ti-chee*. HABITAT, HABITS AND STATUS Regularly forages in mixed species flocks. Uncommon resident in closed woodland, especially Tai Po Kau. (T)

Black-naped Oriole ■ *Oriolus chinensis* 黑枕黃鸝 23–28cm

DESCRIPTION Adult male bright yellow with black facial mask extending back to join at nape. Wings (primaries and outer secondaries) black, and (inner secondaries and tertials) yellow; tips of black primary coverts yellow, forming a small wing-patch. Tail black with yellow 'corners'. Strong pointed bill reddish-pink. Female similar to male, but with yellowish-green mantle and much of wing. Immature has whitish underparts heavily streaked black, yellow vent and green-washed upperparts. Voice: loud, far-carrying, flute-like whistle, *ee lee-oo*. HABITAT, HABITS AND STATUS Common autumn passage migrant, far less numerous in spring, in widespread, open wooded areas mainly in New Territories. Occasional breeding and winter records. (M, L, I)

◾ Drongos ◾

Black Drongo ◾ *Dicrurus macrocercus* 黑卷尾 29.5cm

DESCRIPTION Large, all-black, slim bird with long, deeply forked tail. Plumage has dark blue sheen in bright light. Stout black bill with obvious rictal bristles. Iris deep ruby. Immatures have whitish fringes to underparts and whitish undertail-coverts. **HABITAT, HABITS AND STATUS** Very upright perching stance in prominent positions, from which it sallies to capture insect food. Aggressive towards larger birds such as crows, which it harasses in flight. Common in spring, summer and autumn in widespread, open-wooded, agricultural and waste areas. Less common in winter, when largely recorded in north-west New Territories. (M, L, I)

Ashy Drongo

◾ *Dicrurus leucophaeus* 灰卷尾 29cm

DESCRIPTION Similar size to Black Drongo (see above), with which it shares slim appearance and long, deeply forked tail. Plumage iron grey above and paler grey below. Head shows black feathering around bill-base and whitish patch around eye (race *leucogenis*). Stout black bill and deep ruby iris. **HABITAT, HABITS AND STATUS** Often perches prominently. Uncommon winter visitor to widespread wooded areas. (T)

■ Drongos & Monarchs ■

Hair-crested Drongo ■ *Dicrurus hottentottus* 髮冠卷尾 32cm

DESCRIPTION Similar to Black Drongo (see p. 95), but slightly larger with broader tail, the tips clearly curling outwards and upwards, most readily seen in flight. Crest of long, hair-like feathers. Wings and tail have glossy green sheen. Head, neck, mantle and breast have dark blue glossy tips, giving spangled appearance in strong light. **HABITAT, HABITS AND STATUS** Common resident, passage migrant and winter visitor, favouring more heavily wooded areas than Black Drongo. (T, I)

Black-naped Monarch ■ *Hypothymis azurea* 黑枕王鶲 15–17cm

DESCRIPTION Male azure, deeper-toned on back, with white belly and vent. Small black patch above base of bill, black nuchal crest and narrow gorget across centre of breast. Female duller than male, with blue head, brownish-grey upperparts and pale blue-grey breast. Lacks nuchal crest and black breast-band. **HABITAT, HABITS AND STATUS** Uncommon winter visitor and passage migrant to widespread wooded areas. (T, I)

▪ Monarchs ▪

Amur Paradise-Flycatcher
▪ *Terpsiphone incei* 綬帶 20cm, not including male's tail streamers of up to 30cm

DESCRIPTION Male has glossy, blue-black head and bushy crest. Sky-blue eye-ring, bright rufous-brown upperparts and remarkably long tail (central rectrices), which exceeds body length. Bluish-grey breast merges into white belly and vent. Young males lack tail streamers. Female duller with shorter crest, pale throat and medium-length tail without elongated central rectrices. A white morph occurs in males only and has been recorded in Hong Kong. Hood black, as in rufous morph; rest of plumage white except for black primaries. **HABITAT, HABITS AND STATUS** Uncommon autumn migrant, less frequent in spring, and rare winter visitor to widespread, well-wooded areas. (M, T, I)

Japanese Paradise-Flycatcher
▪ *Terpsiphone atrocaudata* 紫綬帶 20cm, not including male's tail streamers of up to 25cm

DESCRIPTION Sexes similar to Amur Paradise-Flycatcher (see above) but somewhat smaller. Male darker with satin black upperparts and long tail, the former, especially the mantle, showing a purplish gloss. Head, throat and breast black, merging into whitish belly and vent. Conspicuous bright blue eye-ring. Bill blue. No white morph exists. Female as Amur Paradise-Flycatcher, but upperparts and tail darker and browner mostly lacking rufous hues. **HABITAT, HABITS AND STATUS** Uncommon passage migrant to widespread, well-wooded areas. (M, T, I)

CROWS

Azure-winged Magpie
■ *Cyanopica cyanus* 灰喜鵲 36–38cm

DESCRIPTION Distinctive long-tailed, colourful corvid. Head black and throat whitish; remainder of underparts pinkish-buff. Mantle pinkish-buff suffused with pale blue, wings and tail sky-blue, and tip of tail white. HABITAT, HABITS AND STATUS Occurs in light woodland with open areas. Gregarious, typically in small flocks moving through trees, though readily comes to ground to forage. Locally common principally in Mai Po NR, but regularly observed elsewhere in northern New Territories. (M, L)

Red-billed Blue Magpie
■ *Urocissa erythrorhyncha* 紅嘴藍鵲 53–64cm, including tail of up to 47cm

DESCRIPTION Unmistakable large, exotic, exceptionally long-tailed crow. Upperparts and markedly graduated and inward-curving tail bright cobalt-blue, the mantle rather duller grey toned, the tail feathers broadly tipped black and white. Head, throat and upper breast black with white nape-patch. Underparts white. Strong bill and legs bright red. Voice: frequently vocal with wide repertoire, including distinctive *spink-spink*. Readily mimics calls of other species. HABITAT, HABITS AND STATUS Common and widespread resident in open woodland and urban parks throughout Hong Kong. Also inhabits mangroves. Sociable, usually roaming in small (family) parties. (M, L, T, H, I)

■ CROWS ■

Grey Treepie ■ *Dendrocitta formosae* 灰樹鵲 36–40cm

DESCRIPTION Drab, long-tailed corvid. 'Face' and throat sooty-black, nape and collar grey merging to greyish-brown on breast; greyer on belly with chestnut undertail-coverts. Mantle dull brown; wings and tail blackish. Small white patch at base of primaries and whitish rump conspicuous in flight. **HABITAT, HABITS AND STATUS** Typically occurs in small groups. Uncommon resident in widespread wooded areas. (T)

Eurasian Magpie ■ *Pica pica* 喜鵲 44–46cm

DESCRIPTION Large, long-tailed, pied crow. Plumage black glossed purplish-blue and green. White lower breast, belly, flanks and large wing-patch across scapulars and outer primaries. Stout black bill. **HABITAT, HABITS AND STATUS** Common and widespread resident in open, lightly wooded and urban areas throughout Hong Kong. Often builds nest in man-made structures such as light towers and disused crane jibs. (M, L, T, H, I)

CROWS

House Crow
■ *Corvus splendens* 家鴉 40–43cm

DESCRIPTION Smaller than other two black crows resident in Hong Kong. All black with broad brownish-grey band from rear crown and nape to mantle, and extending across breast. **HABITAT, HABITS AND STATUS** Locally common resident deriving from introduced stock. Mainly in urban area of West Kowloon.

Collared Crow ■ *Corvus torquatus* 白頸鴉 50–55cm

DESCRIPTION Large black crow with conspicuous white band on neck, mantle and across breast. Bill less heavy than Large-billed Crow's (see opposite). This is in strong contrast to elsewhere in China where species is becoming increasingly rare. Voice: a rough *haarr* usually uttered in a sequence of two or three notes. Large-billed Crow's call is a shorter, more resonant, oft-repeated *ka*. **HABITAT, HABITS AND STATUS** Locally common resident favouring widespread coastal areas, particularly fish ponds in north-west New Territories. Regular roost in intertidal mangroves at Mai Po NR commonly holds more than 100 birds in both summer and winter. (M, L)

CROWS & FAIRY FLYCATCHER

Large-billed Crow ■ *Corvus macrorynchos* 大嘴烏鴉 46–59cm

DESCRIPTION Large black crow with noticeably heavy bill with strongly curved culmen and steep forehead. The only all-black crow resident in Hong Kong. **HABITAT, HABITS AND STATUS** Common resident in open wooded areas throughout Hong Kong, including urban edges. (M, L, T, H, I)

Grey-headed Canary-flycatcher
■ *Culicicapa ceylonensis* 方尾鶲 12–13cm

DESCRIPTION Small, bright flycatcher with brilliant yellow belly, contrasting grey head and upper breast, and green upperparts. Blackish wings and tail edged bright green. Rear-peaked crown gives slightly crested appearance. Rather large eye with narrow whitish eye-ring. Legs noticeably pink. Upright stance when perched, often flicking tail. Voice: short, dry rattle and quite loud, whistled *ti-wu, ti-wu, wee*. **HABITAT, HABITS AND STATUS** Active and vocal. Tendency to stay lower to the ground than most other flycatchers and less prone to perching for lengthy periods. Uncommon winter visitor to widespread woodland. (T, I)

■ TITS ■

seen in garden Ming Tong

Cinereous Tit
■ *Parus cinereus* 蒼背山雀 14cm

DESCRIPTION Smallish and strongly patterned. Male has black head with large white patch stretching across 'face' below eye level. Chin and throat black, with black extending as a vertical stripe down centre of breast to vent. Upperparts grey; wing-bar created by white tips to greater coverts. Underparts buffy-grey; darker on flanks. Tail grey-black with white outer rectrices. Female similar to male, but with narrower frontal stripe. HABITAT, HABITS AND STATUS Active and acrobatic in search of food. Common resident in woodland of all types, including mangroves. (M, L, T, H, I)

Yellow-cheeked Tit
■ *Machlolophus spilonotus* 黃頰山雀 14–15.5cm

DESCRIPTION Boldly marked black, white, grey and yellow bird. Male has black crown and tufted crest, the latter tipped bright yellow. 'Face' and rear crown bright yellow; black stripe from back of eye across ear-coverts. Black chin, throat and breast, with black continuing as broad central stripe down belly and vent. Rest of underparts dingy-white. Mantle grey, tipped black; wing-coverts tipped white, forming double wing-bar. Tail black, edged white. Female and immature duller than male, without black stripe on underparts, which is replaced by grubby yellowish-olive stripe. HABITAT, HABITS AND STATUS Uncommon resident deriving from captive stock. Restricted to woodland in central New Territories. (T)

▪ Penduline Tit & Lark ▪

Chinese Penduline Tit ▪ *Remiz consobrinus* 中華攀雀 10.5cm

DESCRIPTION Very small. Male has grey crown and white-bordered black, shrike-like mask. Mantle chestnut; rest of upperparts greyish-brown relieved by broad chestnut wing-coverts bar. Underparts greyish-buff. Female duller than male, with less extensive brown mask and whiter throat. Voice: high-pitched, wispy, descending *see-oo*, often given in flight. **HABITAT, HABITS AND STATUS** Common migrant, mostly in autumn, and winter visitor mainly to reed beds in north-west New Territories. (M, L)

Male

First-winter female

Eurasian Skylark ▪ *Alauda arvensis* 雲雀 18–19cm

DESCRIPTION Medium-sized lark with inconspicuous small crest. Upperparts greyish-brown strongly marked with blackish-brown. Breast buff heavily streaked blackish-brown; rest of underparts dull white. Head has broad creamy-white supercilium curving down behind ear-coverts to white throat. Narrow white trailing edge to secondaries and inner primaries. Tail blackish with white outer rectrices. **SIMILAR SPECIES** Scarce **Oriental Skylark** *A. gulgula* distinguished by slightly smaller size, and buff rather than white trailing edge to wing and outer-tail feathers. Bill finer and marginally longer, and tail and primary projection shorter. **HABITAT, HABITS AND STATUS** Uncommon autumn passage migrant and scarce winter visitor to widespread open, grassy areas principally in New Territories, especially Long Valley. (M, L)

regular in garden

BULBULS

Red-whiskered Bulbul ◾ *Pycnonotus jocosus* 紅耳鵯 18–20.5cm

DESCRIPTION Unmistakable handsome bulbul with long, spiky black crest. Black crown, nape and half-collar. White ear-covert patch with red spot above and thin black line below separating patch from white chin, throat and upper breast. Upperparts brown; tail brown with white 'corners'. Underparts grey with bright red vent. Bill and legs black. **HABITAT, HABITS AND STATUS** Abundant resident; one of the most common and widespread birds throughout Hong Kong, including in urban areas with trees. Forms flocks outside breeding season. Commonly perches conspicuously. (M, L, T, H, I)

Chinese Bulbul
◾ *Pycnonotus sinensis* 白頭鵯 19cm

DESCRIPTION Light-coloured bulbul with black head splashed with bold white band broadening from eye to nape. Small white spot on ear-coverts. Upperparts dull olive; flight feathers and outer rectrices broadly fringed yellowish-green. Underparts greyish-buff with paler whitish belly. Bill and legs black. Juvenile lacks head pattern of adult, which is replaced largely by plain olive. **HABITAT, HABITS AND STATUS** Abundant resident, migrant and winter visitor. The most common and widespread bird throughout Hong Kong. Occurs in virtually all habitats, including urban areas. Forms flocks outside breeding season. Noisy and conspicuous. (M, L, T, H, I)

◾ BULBULS ◾

Sooty-headed Bulbul
◾ *Pycnonotus aurigaster* 白喉紅臀鵯 19–21cm

DESCRIPTION Forehead and crown to eye level, chin and sides of 'face' below eye dull sooty-black, forming distinct hood. Short, pointed crest. Upperparts dull brown; rump whitish-grey; tail black with pale buff-white tip. Ear-coverts, throat and underparts grey-buff with red vent. Juvenile has yellow vent. **HABITAT, HABITS AND STATUS** Common, widespread yet localized resident throughout Hong Kong, preferring lightly wooded secondary growth. Forms flocks outside breeding season. (M, L, T, I)

Mountain Bulbul ◾ *Ixos macclellandii* 綠翅短腳鵯 21–24cm

DESCRIPTION Rather large bulbul. Reddish-brown forehead and crown with loose crest. Ear-coverts and sides of neck tinged rufous. Throat grey, heavily streaked white. Mantle and rump grey. Wings and tail olive; outer fringes of primaries and secondaries noticeably yellow. Breast reddish-buff, merging into creamy-buff belly and pale yellow vent. **HABITAT, HABITS AND STATUS** Uncommon resident in woodland of central New Territories. First recorded in 2001 at Tai Po Kau and gradually extending range. Frequently raises crest and puffs out shaggy throat feathers. (T)

■ Bulbuls, Swallows & Martins ■

Chestnut Bulbul
■ *Hemixos castanonotus*
栗背短腳鵯 21.5cm

DESCRIPTION Bright chestnut head and mantle, black crown and loose crest. Blackish-brown wings and tail. Conspicuously white throat. Breast and flanks grey shading into whitish belly and vent. **HABITAT, HABITS AND STATUS** Common resident and winter visitor to widespread wooded areas throughout Hong Kong. Irregular winter irruptions occur. (T, H, I)

Pale Martin ■ *Riparia diluta* 淡色沙燕 12cm

DESCRIPTION Small brown and white martin. Head, upperparts and breast-band greyish-brown. Chin, throat and underparts white. Tail notched. **HABITAT, HABITS AND STATUS** Uncommon passage migrant with infrequent winter records. Widespread in open spaces but mainly seen hawking insects over fish ponds in Deep Bay area. (M, I)

◾ Swallows & Martins ◾

Barn Swallow ◾ *Hirundo rustica* 家燕 17–19 cm

DESCRIPTION Familiar, streamlined aerial passerine. Dark, glossy blue above. Deep red forehead and throat bordered by blue-black breast-band; white or buff-washed (rarely rufous) underparts. Tail distinctly forked with long, elongated outer-tail feathers. **HABITAT, HABITS AND STATUS** Present throughout the year. Abundant passage migrant and common summer breeding visitor. Uncommon in winter. Widespread. Feeds over open land, marshes and fish ponds. Nests on buildings in villages and more densely populated urban areas, constructing nest with mud collected from puddle edges. (M, L, H, U)

seen in Fanling Wai village

Asian House Martin ◾ *Delichon dasypus* 煙腹毛腳燕 13cm

DESCRIPTION Head, upperparts and tail glossy dark blue; conspicuous white rump. White throat and underparts washed grey. Underwing grey with contrasting darker coverts. Tail slightly forked. **SIMILAR SPECIES** Vagrant **Common House Martin** *D. urbicum* distinguished with difficulty by more glossy upperparts, pure white underparts, paler underwing-coverts, longer white rump-patch extending onto uppertail-coverts, and slightly deeper forked tail. **HABITAT, HABITS AND STATUS** Uncommon passage migrant, less frequent in autumn, and rarely in winter. Widespread in open areas, but mainly over fish ponds in Deep Bay area. (M, L)

■ Swallows, Martins & Wren-babbler ■

Red-rumped Swallow ■ *Cecropis daurica* 金腰燕 16–17cm

DESCRIPTION Jizz similar to Barn Swallow's (see p. 107), though tail streamers are even longer and flight is heavier. Cap, mantle, wings and tail glossy blue. Narrow supercilium and sides of nape brownish-orange. Rump buffy-orange of varying intensity. Whitish ground colour to thin black streaking on ear-coverts, throat and underparts. Black patch of undertail-coverts is useful identification feature on birds flying overhead. **HABITAT, HABITS AND STATUS** Common passage migrant and winter visitor with small breeding population. Mainly frequents fish ponds and open ground in Deep Bay area, though recorded from widespread locations throughout Hong Kong. (M, L)

Pygmy Wren-babbler ■ *Pnoepyga pusilla* 小鷦鶥 7.5–9cm

DESCRIPTION Diminutive and 'tail-less'. Upperparts warm brown; neck, mantle and wing-coverts spotted buffy-white. Underparts off-white boldly scaled dark grey-brown. Eyes large; legs pale pinkish. Voice: loud, repeated, high-pitched *tee-tee* with pause between the two notes, or often just single *tee*. **HABITAT, HABITS AND STATUS** Highly terrestrial, foraging in leaf litter on woodland floor. Locally common resident. Expanding range in central and north-east New Territories. (T)

BUSH WARBLERS & ALLIES

Mountain Tailorbird
■ *Phyllergates cuculatus* 金頭縫葉鶯 10–12cm

DESCRIPTION Small, long-tailed – though tail not as long as Common Tailorbird's (see p. 120) – and rather long-billed bird. Bright yellow lower underparts and rump distinctive. Cap brownish-orange, 'face' and neck grey, mantle and wings olive, and throat and breast washed grey. **HABITAT, HABITS AND STATUS** Locally common winter visitor and uncommon resident in widespread wooded areas. First recorded in Hong Kong in 1999, since when it has rapidly expanded its range in central and north-east New Territories. (T, I)

Japanese Bush Warbler ■ *Horornis diphone* 日本樹鶯 15–18cm

DESCRIPTION Rather large, plain, sombre-coloured warbler. Upperparts warm brown. Pale creamy supercilium and dark eye-stripe. Crown and tail dull rufous. Off-white chin and throat, greyish-brown breast and paler belly. Tail rather long. Legs pale pink. Voice: call a single sharp *chup*, often uttered in quick succession. **SIMILAR SPECIES** Recently split **Manchurian Bush Warbler** *H. borealis* is virtually identical and field identification features have yet to be determined. Best distinguished by call, a harsh scolding *rurrt*, very different from Japanese Bush Warbler's. **HABITAT, HABITS AND STATUS** Uncommon but widespread winter visitor and passage migrant, with a preponderance of records from Deep Bay area. Favours dense tangles of undergrowth in shrubby locations. Skulking; presence often first established by voice. (M, I)

▪ Bush Warblers ▪

Brown-flanked Bush Warbler
▪ *Horornis fortipes* 強腳樹鶯 11–12.5cm

DESCRIPTION Very plain. Upperparts warmish-brown with a poorly defined yellow-brown supercilium. Underparts lighter greyish-brown; more olive on flanks. Bill dark with pale base (often extensive) to lower mandible. Legs pale brownish-pink. Voice: song a long, drawn-out, rising *weeee*, followed by an explosive whiplash *wichoo*. **HABITAT, HABITS AND STATUS** Typically skulking in tangles of undergrowth. Locally common and widespread winter visitor increasingly recorded breeding in upland scrub such as upper slopes of Tai Mo Shan. (T, I)

Asian Stubtail ▪ *Urosphena squameiceps* 鱗頭樹鶯 9.5–10.5cm

DESCRIPTION Very small, almost tail-less bird. Warm brown upperparts, striking long, broad creamy supercilium and blackish eye-stripe. Underparts grey-buff; throat and breast paler, merging into grey belly and darker brown flanks and vent. Stubby tail extends just beyond primary tips. Legs pink. **HABITAT, HABITS AND STATUS** Common and widespread winter visitor to mature closed canopy woodland. Inconspicuous, foraging on the ground among leaf litter, where its actions are almost mouse-like. (T, I)

▪ Leaf Warblers ▪

Dusky Warbler ▪ *Phylloscopus fuscatus* 褐柳鶯 11cm

DESCRIPTION Medium-sized leaf warbler entirely lacking green tones. Upperparts greyish-brown. Supercilium white, becoming buff behind eye, and dark brown eye-stripe. Underparts off-white merging to brownish buff on flanks and ventral area. Bill short and fine; upper mandible dark, lower horn at base with dark tip. Voice: call a sharp, staccato *chack* recalling Stejneger's Stonechat (see p. 144). **SIMILAR SPECIES** See Radde's Warbler (below). **HABITAT, HABITS AND STATUS** Abundant and widespread winter visitor and passage migrant. Favours low vegetation often close to water. Unobtrusive but active, remaining close to the ground. Most frequently located by call. (M, L, T, H, I)

Radde's Warbler ▪ *Phylloscopus schwarzi* 巨嘴柳鶯 12.5cm

DESCRIPTION Like a bright Dusky Warbler (see above), often with bronze tint to upperparts and hint of yellow on underparts. Separated by slightly larger size, comparatively large head, and long pale supercilium that is broader and buffier in front of eye, and especially by short, stout all-pale bill. Legs pink and relatively stout, and feet rather large. Voice: call an agitated *chep* often repeated in a quick uneven series. Softer than Dusky Warbler's call. **SIMILAR SPECIES** In addition to Dusky Warbler, might be confused with rare **Yellow-streaked Warbler** *P. armandii*, which has an indistinctly streaked throat, longer, sharper bill and thinner legs. **HABITAT, HABITS AND STATUS** Uncommon autumn migrant and rare winter visitor to widespread areas of scrub and woodland throughout Hong Kong. Skulking, foraging movements somewhat slow and deliberate at low level in shrubbery and on the ground. (M, T)

■ LEAF WARBLERS ■

Pallas's Leaf Warbler ■ *Phylloscopus proregulus* 黃腰柳鶯 9cm

DESCRIPTION Diminutive, bright leaf warbler. Head boldly patterned: pale lemon crown-stripe; yellowish-olive brow; bright yellow, broad and long supercilium fading first to lemon, then white behind eye; broad suffused black eye-stripe; narrow pale crescent below eye and pale greyish-lemon cheeks. Mantle bright olive-green. Grey wing-coverts broadly tipped lemon to produce broad, double wing-bars. Flight feathers edged yellow. Rump pale lemon; tail yellowish-green. Bill black with pinkish base to lower mandible. Legs pinkish-brown. **HABITAT, HABITS AND STATUS** Common and widespread winter visitor and passage migrant to woodland and wooded areas. Very active; frequently hovers when feeding, when lemon rump is conspicuous. (M, L, T, H, I)

Yellow-browed Warbler ■ *Phylloscopus inornatus* 黃眉柳鶯 10cm

DESCRIPTION Small; only slightly larger than Pallas's Leaf Warbler (see above). Greyish-green upperparts, broad, long creamy supercilium and black eye-stripe. Sometimes shows subdued pale central crown-stripe. Pronounced creamy to white broad, double wing-bars produced by pale tips to median and greater wing-coverts, the latter wider and more prominent, contrasting with blackish bases to secondaries. Tertials also broadly edged cream to white. Underparts off-white. Bill fine and dark with pinkish base to lower mandible. Legs pinkish-brown. Voice: call note a loud, rising *tsuweest*. **SIMILAR SPECIES** Vagrant **Hume's Leaf Warbler** *P. humei* greyer, usually with dark legs and shorter, lower pitched call. **HABITAT, HABITS AND STATUS** Abundant, widespread winter visitor and passage migrant to areas of woodland and scattered trees throughout Hong Kong. Active; commonly flicks wings. (M, L, T, H, I)

First-winter male

LEAF WARBLERS

Arctic Warbler ■ *Phylloscopus borealis* 極北柳鶯 12–13cm

DESCRIPTION Largish, long-winged leaf warbler. Upperparts dull olive-green, long and conspicuous pale yellow supercilium from lores (not bill) to nape, and black eye-stripe from bill-base. Narrow creamy to white wing-bar formed by pale tips to greater coverts. Sometimes shows a second diffuse wing-bar on median coverts. Bright lemon-yellow alula often visible. Underparts greyish-white. Bill longish and dark, with bright fleshy-orange base to lower mandible. Legs orangey-brown. Voice: distinctive monosyllabic *dzick*. **SIMILAR SPECIES** Rare **Japanese Leaf Warbler** *P. xanthodryas* possibly indistinguishable in the field, but may show brighter green upperparts and yellower flanks and wing-bar. Two-barred Warbler (see below) has supercilia meeting at forehead, darker legs and sparrow-like call. **HABITAT, HABITS AND STATUS** Passage migrant, common in autumn, far less numerous in spring, to widespread wooded areas and shrubbery throughout Hong Kong. (M, T, H, I)

Two-barred Warbler ■ *Phylloscopus plumbeitarsus* 雙斑柳鶯 11.5–12cm

DESCRIPTION Medium-sized leaf warbler with narrow whitish double wing-bars. Upperparts dark green; long creamy supercilium broadening behind eye. Underparts off-white. Bill has dark upper mandible and orangey lower mandible. Legs bluish-grey to warm blackish-brown. Voice: sparrow like *chee-wee* or *chee-le-wee*. **HABITAT, HABITS AND STATUS** Uncommon passage migrant, far less numerous in spring than autumn, and winter visitor to widespread woodland and wooded areas. (M, T, I)

▪ LEAF WARBLERS ▪

Pale-legged Leaf Warbler ▪ *Phylloscopus tenellipes* 淡腳柳鶯 10–11cm

DESCRIPTION Smallish leaf warbler with olive-green or olive-brown upperparts. Conspicuously broad and very long whitish supercilium underlined with broad blackish eye-stripe. Faint double wing-bar not always apparent. Underparts off-white. Bill pinkish-brown with dark culmen. Legs conspicuously pale flesh. Voice: metallic *chuie* or *chi*. **SIMILAR SPECIES** Scarcer migrant **Sakhalin Leaf Warbler** *P. borealoides* doubtfully separable in the field. **HABITAT, HABITS AND STATUS** Uncommon autumn passage migrant and scarce spring and winter visitor to widespread thickly wooded areas. Usually keeps low in undergrowth, inconspicuously feeding on or near the ground. (M, T, I)

Eastern Crowned Warbler ▪ *Phylloscopus coronatus* 冕柳鶯 11–12cm

DESCRIPTION Largish leaf warbler. Strongly marked head with wide, creamy-green central crown-stripe bordered by blackish brow and very long, off-white supercilium from bill-base almost to nape. Blackish eye-stripe. Upperparts greenish-olive; flight feathers with yellow-green edges. Often shows a narrow whitish wing-bar on greater coverts. Underparts white with lemon-yellow undertail-coverts. Bill has dark upper mandible, and dull orange lower mandible. Legs pinkish-grey. **HABITAT, HABITS AND STATUS** Uncommon autumn passage migrant, less numerous in spring and rare winter visitor, to widespread woodland areas. (T, I)

LEAF WARBLERS & REED WARBLERS

Goodson's Leaf Warbler ■ *Phylloscopus goodsoni* 古氏[冠紋]柳鶯
11–12cm

DESCRIPTION Medium-sized, rather bright leaf warbler with boldly patterned head. Crown dark olive-green with broad yellowish-olive central stripe. Long, broad greenish-yellow supercilium extends almost to nape; dark olive eye-stripe. Upperparts olive-green with pale yellow, wide double wing-bars. Throat and breast suffused yellow. Flanks and belly greyish. Two outer rectrices edged white. Bill distinctly bicoloured; upper mandible blackish, lower mandible bright pinkish-orange. Leg colour varies from dark grey to yellowish. **HABITAT, HABITS AND STATUS** Habitually searches tree boughs, utilizing nuthatch-like feeding method. Locally common winter visitor to widespread woodland areas. (T, I)

Oriental Reed Warbler ■ *Acrocephalus orientalis* 東方大葦鶯 18.5cm

DESCRIPTION Large, robust reed warbler. Upperparts dull grey-brown with richer brown rump; tail shows white tips to outer rectrices. Pale supercilium above blackish eye-stripe. Underparts off-white, whiter on throat, with fine brown streaking on breast and warmer toned buff flanks and vent. Legs grey. **SIMILAR SPECIES** Scarce **Thick-billed Warbler** *Iduna aedon* lacks supercilium and eye-stripe, and has shortish stout bill and shorter primary projection. **HABITAT, HABITS AND STATUS** Common passage migrant and occasional summer and winter visitor to widespread wetland and grassy areas throughout Hong Kong, especially reed beds and fish ponds around Deep Bay. (M, L)

▪ Reed Warblers & Grass Warblers ▪

Black-browed Reed Warbler
■ *Acrocephalus bistrigiceps* 黑眉葦鶯 13.5cm

DESCRIPTION Rather small reed warbler. Warm brown above with rufous blush on rump. Bold head pattern dominated by broad creamy-white supercilium, bordered above by distinctive black brow and below by narrow black eye-stripe. Chin and throat white. Underparts off-white with rufous-buff flush to sides of breast and flanks. **HABITAT, HABITS AND STATUS** Locally common passage migrant and scarce winter visitor mainly to reed beds of Deep Bay area with occasional records elsewhere. Tends to remain deep in vegetation, although in spring ascends to reed tops to sing. (M, L)

Russet Bush Warbler ■ *Locustella mandelli* 高山短翅鶯 13cm

DESCRIPTION Upperparts wholly dark reddish-brown. Indistinct pale, rather short supercilium and bolder eye-ring. Throat creamy-white; rest of underparts dull brown, the undertail-coverts tipped buff and forming pale crescents. Voice: call a repeated stony *chick* or *ch-ick*; song a repeated harsh, insect-like *zee-bit*. **HABITAT, HABITS AND STATUS** Inveterate skulker in low, dense tangles. Uncommon winter visitor to widespread scrub and grassy areas, and rare breeder on uplands. (I)

▪ GRASS WARBLERS ▪

Lanceolated Warbler ▪ *Locustella lanceolata* 矛斑蝗鶯 12–12.5cm

DESCRIPTION Smallest, most streaked *Locustella*. Upperparts' ground colour pale grey-brown. Underparts lighter and head conspicuously finely streaked blackish-brown; mantle, rump, throat, breast, flanks and undertail-coverts display larger streaking. Pale, indistinct supercilium. **HABITAT, HABITS AND STATUS** Extreme skulker in low vegetation, especially in moist areas. Uncommon migrant, more numerous in autumn, with occasional winter records. Most records from Deep Bay area. (M, L)

Pallas's Grasshopper Warbler ▪ *Locustella certhiola* 小蝗鶯 13.5cm

DESCRIPTION Remarkably variable in both plumage tones and size. Extreme individuals different enough to suggest two species. Most common plumage rather bright with tawny ground colour to upperparts and more rusty-yellow (although can be off-white) underparts. Fairly conspicuous yellowish-brown supercilium. Crown heavily streaked blackish-grey; mantle streaked in parallel lines; scapulars, median and greater coverts, and tertials black, fringed rufous. Rump plainer and distinctly rufous. Breast usually sparsely streaked, although can show almost no streaking. White tips to dark tail feathers (not central rectrices) diagnostic. **HABITAT, HABITS AND STATUS** Skulking nature in low, moist grassland and reed beds makes species very difficult to see well. Common autumn migrant, scarce in spring and winter. Most records from Deep Bay area. (M, L)

CISTICOLAS & ALLIES

Zitting Cisticola
■ *Cisticola juncidis* 棕扇尾鶯 10cm

DESCRIPTION Diminutive. Sexes are similar. Upperparts brown streaked black. Crown with rufescent tinge streaked black. Rump warm rufous-brown; tail brown with black subterminal band and white tips. Whitish supercilium, and white throat and underparts with buff flanks. Voice: call a loud, staccato repeated *zit zit zit*. Song similar, longer and more regular, delivered in high, bouncing, circular display flight. **HABITAT, HABITS AND STATUS** Tail often fanned in flight. Common and widespread passage migrant and winter visitor to open grasslands in New Territories. Breeds in Deep Bay area. (M, L, I)

Golden-headed Cisticola ■ *Cisticola exilis* 金頭扇尾鶯 9cm

DESCRIPTION Male in breeding plumage, which is attained in Hong Kong in early spring, has unstreaked rufous-brown head and brown upperparts with blackish streaks on mantle. Wing feathers black edged buff-brown; tail blackish-brown with narrow rufous-buff edges and buff tips. Throat and central breast white; rest of underparts orangey-buff. In winter more streaked on crown and rump, with much longer tail, risking confusion with prinias. Female marginally smaller, duller and more streaked than male, including on crown in both breeding and non-breeding plumages. Voice: call a repeated, wheezy *wheea*. **SIMILAR SPECIES** See Zitting Cisticola (above). Separated in winter by buffer supercilium and nape, greyer ear-coverts, much longer tail with duller tips and call. **HABITAT, HABITS AND STATUS** Locally common winter visitor to widespread grassy regions throughout Hong Kong. (L, I)

▪ Cisticolas & Allies ▪

Yellow-bellied Prinia
▪ *Prinia flaviventris* 黃腹鷦鶯 11cm

DESCRIPTION Rather colourful, long-tailed bird with a delicate structure. Greyish-brown upperparts, lavender-grey head, and white loral stripe and eye-ring. White chin, throat and neck-sides; pale yellow underparts. Juvenile has brown head and buffier flanks. **HABITAT, HABITS AND STATUS** Long tail often cocked, switched and even rotated. Abundant and widespread resident of open grass and wetland areas throughout Hong Kong. (M, L, I)

Plain Prinia ▪ *Prinia inornata* 純色鷦鶯 11cm

DESCRIPTION Small, drab-coloured bird with plain greyish-brown upperparts, and pale buffish supercilium, lores and ear-coverts. Underparts pale yellowish-buff. Tail long and graduated with white tips. Eye reddish-brown, legs pink. Voice: varied including a buzzy trill, a loud mewing *eeee* and a high-pitched repetitive *tee-tee-tee-tee-tee*.
SIMILAR SPECIES Juvenile Yellow-bellied Prinia (see above) has finer bill and shorter tail. **HABITAT, HABITS AND STATUS** Locally common and widespread resident in open grass and wetland, especially north and east New Territories. (M, L, I)

CISTICOLAS, ALLIES & OLD WORLD BABBLERS

Common Tailorbird ■ *Orthotomus sutorius* 長尾縫葉鶯 11–14cm

DESCRIPTION Small bird with bright greyish-green upperparts and orangey-brown cap. Broad greyish supercilium and grey 'face'. Chin and underparts white, merging to grey on breast-sides and flanks. Long, graduated tail habitually held cocked. Female shorter tailed than male, with streaking on 'face' and grey-buff breast. Bill rather long and slender; dark upper mandible, pinkish lower mandible. Voice: call a loud *chee chee chee chee…*; song a loud, urgent, repetitive *chiu chiu chiu chiu…* HABITAT, HABITS AND STATUS Keeps to lower level of vegetation. Common and widespread resident throughout Hong Kong wherever there is bushy cover. (M, L, T, H, I)

Streak-breasted Scimitar Babbler
■ *Pomatorhinus ruficollis* 棕頸鉤嘴 16–19cm

DESCRIPTION Strikingly patterned head. Crown dark olive-brown; long, bold white supercilium curving down at nape; blackish lores and ear-coverts. Nape and sides of neck rufous, upperparts warm brown and tail blackish-brown. Throat and breast white, the latter heavily streaked warm brown; belly and ventral area brown. Bill rather long, decurved and pale horn with dark culmen. Voice: contact call a throaty, rasping *ker ker*; song a repeated clear, three-note whistled *per-per-wer*. HABITAT, HABITS AND STATUS Naturalized resident of captive origin. Roams in small parties, often joining bird waves; otherwise skulking and difficult to observe. Best located by voice. Locally common and widespread resident in thick, bushy hillsides and woodland. (T)

▪ Old World Babblers & Fulvetta ▪

Rufous-capped Babbler
▪ *Stachyridopsis ruficeps* 紅頭穗鶥 12cm

DESCRIPTION Small, yellowish-green babbler. Crown brownish-orange, and face and throat bright yellowish-green. Upperparts olive; wings darker; underparts lighter olive. Bill stout yet pointed and black with pale base to lower mandible. Voice: song a series of six whistled notes, *wu wu wu-wu-wu-wu*, the last four in quicker succession than the first two. **HABITAT, HABITS AND STATUS** Skulking; best located by call. Often in small groups. Common resident, derived from introduced stock, in dense scrub and woodland in central New Territories. (T)

Huet's Fulvetta ▪ *Alcippe hueti* 黑眉雀鶥 12.5–14cm

DESCRIPTION Small, large-headed and 'neckless' bird. Head grey; whitish ring surrounds large eye and long, indistinct, narrow blackish supercilium extends onto neck-sides. Upperparts brown; wings and tail with a grey wash. Breast and flanks orangey-buff; rest of underparts sullied white. **HABITAT, HABITS AND STATUS** Typically in small parties. Uncommon resident, derived from captive stock, in woodland in central New Territories. (T)

▪ Laughingthrushes ▪

Chinese Hwamei ▪ *Garrulax canorus* 畫眉 21–24cm

DESCRIPTION Warm gingery-brown laughingthrush with fine blackish streaking to head, neck and flanks, and conspicuous wide white eye-ring extending back as a line to rear of 'face'. Belly grey. Bill and legs yellow-brown. **HABITAT, HABITS AND STATUS** Common and widespread inhabitant of shrubbery throughout Hong Kong. Unobtrusive. Largely solitary or in pairs. A popular cage bird because of its rich, varied song. (T, I)

Masked Laughingthrush ▪ *Garrulax perspicillatus* 黑臉噪鶥 28–31.5cm

DESCRIPTION Typical laughingthrush with short, rounded wings and long tail. Head, neck and breast grey with wide black mask from forehead to ear-coverts. Upperparts grey-brown; tail somewhat brighter. Underparts paler greyish-buff with deep apricot-buff vent. Voice: call a repetitive, loud, emphatic *pee-u*. **HABITAT, HABITS AND STATUS** Abundant resident in shrubbery throughout Hong Kong, including in urban areas. Sociable; habitually roams in small parties, foraging on the ground. Keeps to lower levels of foliage. Flight characterized by rapid wingbeats interspersed with comparatively lengthy glides. Noisy. (M, L, T, H, I)

▪ LAUGHINGTHRUSHES ▪

Greater Necklaced Laughingthrush ▪ *Garrulax pectoralis*
黑領噪鶥 26.5–34.5cm

DESCRIPTION Large woodland laughingthrush. Crown and upperparts warm brown. 'Face' white boldly streaked black, with black stripe from behind eye. Nuchal collar rufous. Throat white with wide black and grey border at sides extending onto sides of breast as incomplete gorget. Underparts buffy-white with orangey-buff flanks. Prominent black and orange-buff tips to outer-tail feathers.
HABITAT, HABITS AND STATUS Locally common resident, derived from introduced stock, in widespread woodland throughout Hong Kong. (T)

Black-throated Laughingthrush
▪ *Garrulax chinensis* 黑喉噪鶥 23–30cm

DESCRIPTION Dull-coloured laughingthrush with crown, nape and breast dark battleship-grey. Forehead, lores, rear of eye, chin and throat black, with prominent white patch from sides of throat to ear-coverts. Rest of upperparts and underparts dark grey-brown. Voice: very rich, fluty and varied song. Also mimics other species.
HABITAT, HABITS AND STATUS Naturalized resident of captive origin. Locally common and widespread in woodland on Hong Kong Island and New Territories. Largely solitary or in pairs. (T, H)

LAUGHINGTHRUSHES

White-browed Laughingthrush ■ *Garrulax sannio*
白頰噪鶥 22–24cm

DESCRIPTION Smallish laughingthrush with chocolate-brown head and throat. Wide white supercilium joins white cheek-patch in front of eye. Rest of body greyish-brown with rufous wash to outer-tail feathers. **HABITAT, HABITS AND STATUS** Naturalized resident of captive origin. Uncommon inhabitant of thick shrubbery in New Territories. (L)

Blue-winged Minla ■ *Minla cyanouroptera* 藍翅希鶥 14–15.5cm

DESCRIPTION Slim and longish-tailed laughingthrush. Head, neck and underparts grey. Crown streaked bluish-black, and narrow blackish eyebrow broadens towards nape above whitish supercilium and eye-patch. Upperparts buffish-brown, wings blue, and secondaries and tertials black with pale edges. Tail blue merging to black towards tip. **HABITAT, HABITS AND STATUS** Naturalized resident of captive origin. Locally common in woodland mainly in central and north-east New Territories. (T)

LAUGHINGTHRUSHES

Silver-eared Mesia ■ *Leiothrix argentauris* 銀耳相思鳥 15.5–17cm

DESCRIPTION Distinctively multicoloured laughingthrush. Head black; forehead and chin bright yellow; ear-coverts patch silver-grey. Throat, breast, collar and rear neck vermilion. Body olive, paler and more yellow washed on underparts. Wing-coverts vermilion, primaries and secondaries edged orangey-yellow and tertials grey. Rump vermilion, and tail grey edged pale yellow. Bill bright yellow; legs brownish-yellow. **HABITAT, HABITS AND STATUS** Naturalized resident of captive origin. Locally common in thickets and woodland mainly in central New Territories, but also elsewhere, including on Hong Kong Island. (T)

Red-billed Leiothrix ■ *Leiothrix lutea* 紅嘴相思鳥 14–15cm

DESCRIPTION Smallish, multicoloured laughingthrush with greenish-grey upperparts and grey flanks. Lores and wide eye-ring pale yellow; chin and throat yellow, bordered by orange breast-band. Belly and vent pale yellow. Closed wing shows reddish-orange bar above yellow-edged blackish primaries. Notched tail blackish. Stubby, thick bill bright red with black base. **HABITAT, HABITS AND STATUS** Naturalized resident of captive origin. Uncommon and localized in thickets and woodland mainly in central New Territories. (T)

WHITE-EYES & ALLIES

Chestnut-collared Yuhina
■ *Yuhina torqueola* 栗耳鳳鶥 14–15cm

DESCRIPTION Prominent grey crest, grey head and narrow pale eye-ring. Chestnut ear-coverts streaked white and wide rufous collar streaked pale buff. Grey-brown upperparts; darker tail with white-fringed outer rectrices. Throat white and rest of underparts off-white. **HABITAT, HABITS AND STATUS** Generally uncommon winter visitor to wooded areas principally in central New Territories. Usually in noisy flocks. Subject to periodic irruptions, when numerous throughout Hong Kong. Occasional summer records. (T)

Japanese White-eye ■ *Zosterops japonicus* 暗綠繡眼鳥 10–11.5cm

DESCRIPTION Small bird with bright green head and upperparts, sunshine-yellow chin, throat, breast and vent, and grey-white belly. Head dominated by wide, shining white eye-ring and black lores, giving bird a startled appearance. Voice: call a high-pitched, repetitive *chee*. **HABITAT, HABITS AND STATUS** Abundant resident throughout Hong Kong wherever there are trees and shrubs. Gregarious; feeding parties active and very vocal as they move through the trees. (M, L, T, H, I)

Nuthatch & Starlings

Velvet-fronted Nuthatch ■ *Sitta frontalis* 絨頷鳾 12.5cm

DESCRIPTION Short-tailed, 'neckless' and plump bird. Bright blue crown and upperparts. Black forehead-patch, narrow black stripe behind eye and bright yellow iris. Throat white; rest of underparts lightly washed lavender-grey. Longish pointed bill brilliant red. Female lacks post-ocular stripe and has more cinnamon-tinged underparts than male. Juvenile similar to female, but less bright and with blackish bill. **HABITAT, HABITS AND STATUS** Habitually climbs acrobatically on tree trunks and boughs in search of food. Naturalized resident of captive origin. Common in woodland of central New Territories. (T)

Crested Myna ■ *Acridotheres cristatellus* 八哥 25cm

DESCRIPTION Largish myna that is all black except for white tail-tip and small patch on closed wing. Short, stubby crest rises from forehead. Bill cream with pale pink base to lower mandible, legs dull yellow and iris yellow. **HABITAT, HABITS AND STATUS** Abundant resident in open country, extending to urban areas. (M, L, T, H, I)

STARLINGS

Common Myna
- *Acridotheres tristis*
家八哥 25cm

DESCRIPTION Rather large myna with brownish-black head, breast, wings and tail, dull brown mantle and underparts, and white vent and undertail-coverts. Small white patch on closed wing becomes conspicuous in flight. Bill, bare skin around eye and legs yellow. **HABITAT, HABITS AND STATUS** Established from escaped cage birds as a locally common resident in open country in north-west New Territories. (M, L)

Red-billed Starling
- *Spodiopsar sericeus* 絲光椋鳥 22cm

DESCRIPTION Male has strikingly creamy-white head, washed buffy on crown and cheeks, white throat and breast, grey mantle and underparts, and black wings and tail. Small white patch on coverts obvious in flight. Bill and legs bright reddish-orange, the former with a black tip. Female rather dull brown-buff, darker on crown and nape, with wings and tail like male's. **HABITAT, HABITS AND STATUS** Abundant winter visitor to open habitat mainly in north-west New Territories. Small numbers remain in summer to breed. Forms large flocks in winter. (M, L, I)

STARLINGS

White-cheeked Starling
■ *Spodiopsar cineraceus* 灰椋鳥 22cm

DESCRIPTION Rather dark starling. Head and breast sooty-black. Upperparts including wings and tail dark greyish-brown. Rump white, underparts dull greyish-brown, and vent and tail-tip white. Most distinctive features are streaked white forehead and cheeks. Legs and bill yellowish-orange, the latter tipped black. **HABITAT, HABITS AND STATUS** Locally common winter visitor to open country mainly in north-west New Territories, with small numbers remaining in summer to breed. (M, L)

Black-collared Starling ■ *Gracupica nigricollis* 黑領椋鳥 28cm

DESCRIPTION Large, broad-winged, black and white starling. Head white, with broad, bright yellow patch of bare skin extending from bill under eye to ear-coverts. Wide black collar surrounds neck and breast. Upperparts and tail brownish-black, with tips and outer rectrices white; rump and underparts white. Bold white fringes to wing-coverts, secondaries and tertials. Voice: loud, ringing *churr, churr*. **HABITAT, HABITS AND STATUS** Common and widespread resident in open country and urban areas with trees. Often seen in pairs. Perches conspicuously. Noisy. (M, L, T, H, I)

North DISTRICT PARK + Wang Leung Rd

▪ STARLINGS ▪

Daurian Starling ▪ *Agropsar sturninus* 北椋鳥 17cm

DESCRIPTION Rather small, dark starling. Male's head and underparts pale grey; black nape-patch and blackish upperparts glossed purple; glossed green on wings and tail. Broad, creamy bar on scapulars, creamy tips to median and greater coverts and tertials, and buff panel on flight feathers. Rump, vent and undertail-coverts pale buff. Female duller than male, with brown head and upperparts. **HABITAT, HABITS AND STATUS** Uncommon autumn passage migrant, rare in winter and spring, to open country mainly in north-west New Territories. Usually found in association with other starling species. (L)

White-shouldered Starling ▪ *Sturnia sinensis* 灰背椋鳥 17cm

DESCRIPTION Smallish, rather pale starling. Male has whitish-grey head, and pale grey breast, mantle and lower back. Rump and uppertail-coverts whitish. Striking large white wing-patch on scapulars and coverts contrasts with black wings and tail, the latter tipped white. Belly and vent white. Female duller than male, with grey and black plumage tending towards brownish and white tail-tip reduced to corners. Juvenile like female but lacks wing-patch. Bill bluish-grey; iris bluish-white, giving startled appearance. **HABITAT, HABITS AND STATUS** Locally common passage migrant and breeding visitor, scarce in winter. Inhabits open country mainly in north-west New Territories. (M, L, I)

▪ THRUSHES ▪

White's Thrush
▪ *Zoothera aurea* 懷氏地鶇 27cm

DESCRIPTION Large, bulky thrush. Upperparts golden-buff, heavily scalloped black. Underparts paler buff tending to white on flanks, belly and vent, but equally scalloped black except in centre of belly and vent. White 'corners' to tail conspicuous in flight. Underwing boldly striped black and white. **HABITAT, HABITS AND STATUS** Uncommon winter visitor to woodland glades, open, grassy areas and lawns in parks, large gardens and sports grounds throughout Hong Kong. Usually solitary and shy. (L, T, I)

Grey-backed Thrush ▪ *Turdus hortulorum* 灰背鶇 20–23cm

DESCRIPTION Male has lavender-grey head, breast and upperparts, darker on wings, paler on breast. Throat paler with faint streaking, dull orange lower breast and flanks, and white belly and vent. Narrow eye-ring, bill and legs orangey-yellow. Female has greyish-brown head and upperparts with darker wings and tail. Underparts off-white; necklace of blackish spots and dull orange lower breast and flanks. Bill and legs dull yellow. First-winter male similar to female but has greyer upperparts. **HABITAT, HABITS AND STATUS** Common winter visitor and migrant to open wooded and shrubby areas throughout Hong Kong. Forages in leaf litter like other *Turdus* thrushes. (M, L, T, H, I)

THRUSHES

Japanese Thrush
■ *Turdus cardis* 烏灰鶇 21–22cm

DESCRIPTION Male strikingly black and white. Head, upperparts, tail and breast black; rest of underparts white boldly spotted black. Eye-ring, bill and legs orangey-yellow. Female and first-winter birds have warm brown upperparts, conspicuously spotted black, off-white underparts and lightly washed orange flanks. **HABITAT, HABITS AND STATUS** Common winter visitor and passage migrant to widespread wooded open, grassy areas and lawns throughout Hong Kong. (L, T, I)

Chinese Blackbird ■ *Turdus mandarinus* 中國烏鶇 28–29cm

DESCRIPTION Male completely sooty-black with conspicuous orange-yellow bill and legs. Female and first-winter birds blackish-brown with paler chin and throat, often with dark streaks, and/or rufous hue on breast. Eye-ring and bill dull brownish-yellow, the latter sometimes blackish; legs blackish. **HABITAT, HABITS AND STATUS** Common winter visitor and migrant; rare resident. Occurs widely in open woodland and parks throughout Hong Kong. Sometimes in loose flocks. (M, L, T, H, I)

▪ Thrushes ▪

Eyebrowed Thrush ▪ *Turdus obscurus* 白眉鶇 23cm

DESCRIPTION Head pattern distinctive. Adult male has grey head, neck and throat with bold white supercilium, crescent beneath eye joined by line from bill and narrow malar. Upperparts warm brown; underparts brownish-orange with white central belly and vent. Bill has blackish tip, dark brown upper mandible and yellow lower mandible. Legs brownish-yellow. Female duller than male with greyish-brown head, dark malar, and white chin and throat. First-winter male similar to female, but has greyer head and off-white chin and throat. **HABITAT, HABITS AND STATUS** Uncommon passage migrant and scarce winter visitor to widespread wooded areas throughout Hong Kong. (M, T, I)

Pale Thrush ▪ *Turdus pallidus* 白腹鶇 22–23cm

DESCRIPTION Rather plain, drab thrush. Male has dark grey head and throat, warm brown upperparts, greyish-buff primaries, greyish-white underparts with buff flanks, and whitish vent and undertail-coverts. Tail grey with white 'corners' usually obvious in flight. Narrow yellow eye-ring, blackish bill with prominent yellow base to lower mandible and yellowish-brown legs. Female duller than male with lightly streaked whitish throat and dark malar. First-winter birds similar to female, but with pale tips to greater coverts. **HABITAT, HABITS AND STATUS** Common winter visitor and passage migrant to widespread wooded areas throughout Hong Kong. Usually solitary. (L, T, H, I)

Magpie Robins & Old World Flycatchers

Oriental Magpie Robin ■ *Copsychus saularis* 鵲鴝 19–21cm

DESCRIPTION Unmistakable long-tailed, pied robin. Male has glossy black head, upperparts and breast, and tail with conspicuous white outer rectrices. Underparts and long wing-patch white. Female's plumage less contrasting than male's, with grey-toned head, breast and mantle and buff-washed flanks. **HABITAT, HABITS AND STATUS** Abundant resident. Widespread throughout Hong Kong in woodland, shrubbery, urban parks and gardens. Habitually holds tail cocked and forages on the ground. An accomplished songster. (M, L, T, H, I)

Female Mung Tak Garden *Male*

Grey-streaked Flycatcher ■ *Muscicapa griseisticta* 灰紋鶲 12.5–14cm

DESCRIPTION Medium-sized, large-headed brown flycatcher. Upperparts darkish brown, pale lores and narrow eye-ring. White chin, throat and submoustachial that reaches sides of neck. Dark malar stripe extends to dark brown streaking across breast and flanks. Rest of underparts off-white. Longer-winged than Dark-sided or Asian Brown Flycatchers (see opposite), with wing-tips extending well down tail. Bill dark. **HABITAT, HABITS AND STATUS** Noticeably upright stance when perched. Sallies from and returns to perch when fly catching. Uncommon though widespread passage migrant, scarcer in autumn, to wooded areas throughout Hong Kong. (M, T, I)

◾ Old World Flycatchers ◾

Dark-sided Flycatcher ◾ Muscicapa sibirica 烏鶲 13–14cm

DESCRIPTION Medium-sized, large-headed brown flycatcher. Upperparts darkish brown. Pale lores and eye-ring, the former less distinct than in Grey-streaked and Asian Brown Flycatchers (see opposite and below). Dark malar stripe stands out from white chin. White half-collar contrasts with variable dark brown breast and flanks. Rest of underparts off-white. Bill mostly black with limited lighter base to lower mandible. First-winter birds have more prominent white tips to greater coverts forming thin wing-bar, and chequered dark brown and white breast and flanks. **SIMILAR SPECIES** See Grey-streaked and Asian Brown Flycatchers. **HABITAT, HABITS AND STATUS** Uncommon autumn migrant, rare in spring, to widespread wooded areas throughout Hong Kong. Habits similar to those of Grey-streaked and Asian Brown Flycatchers. (T, I)

Asian Brown Flycatcher ◾ Muscicapa latirostris 北灰鶲 12–14cm

DESCRIPTION Medium-sized, large-headed brown flycatcher. Head and upperparts grey-brown, bold pale eye-ring, pale lores and subtle dark malar stripe. Pale fringes and tips to greater coverts, secondaries and tertials. Underparts lightly washed grey-brown with whiter throat. Bill dark with prominent flesh-coloured base and sides to lower mandible. First-winter birds have broader white tips to greater coverts, producing wing-bar of varying prominence. **HABITAT, HABITS AND STATUS** Common winter visitor and passage migrant. Like congeners, sits upright and motionless on exposed perch from which it sallies when catching insect food. (M, L, T, H, I)

Old World Flycatchers

Ferruginous Flycatcher
■ *Muscicapa ferruginea* 棕尾褐鶲 12–13cm

DESCRIPTION Small, rather dark, rufous-brown flycatcher. Head dark grey with conspicuous white eye-ring, more pronounced at rear. Rufous-brown upperparts, brighter rufous-orange lower back and rump, blackish-brown wing-coverts and tertials fringed rufous. Black primaries and blackish-brown tail. Buffish throat, and indistinct, pale, narrow collar below throat extends to sides of neck. Underparts deep rusty-orange. **HABITAT, HABITS AND STATUS** Uncommon spring passage migrant, rare in autumn in widespread wooded areas. (T, I)

Hainan Blue Flycatcher ■ *Cyornis hainanus* 海南藍仙鶲 13–14cm

DESCRIPTION Male has deep blue head, upperparts and breast. Forehead and lesser coverts lighter shiny blue, lores and chin black, throat blackish-blue and belly off-white merging to blue of breast. Female has olive-brown head and upperparts, and reddish-brown rump, tail and fringes to tertials and secondaries. Buffy lores and eye-ring; orangey-buff wash to chin and throat, brighter on breast. Belly, flanks and undertail-coverts white. First-winter male has warm brown head and mantle, broad white line down throat and blue wings (with blackish flight feathers), rump and tail. **SIMILAR SPECIES** Blue-and-white Flycatcher (see opposite) is larger, and male has black throat and breast sharply separated from white belly. Female has pale greyish-brown breast lacking orange tones of Hainan Blue. **HABITAT, HABITS AND STATUS** Common breeding visitor and passage migrant, scarce in winter. Widespread in woodland. (T, I)

Female

Male

◾ OLD WORLD FLYCATCHERS ◾

Blue-and-white Flycatcher
◾ *Cyanoptila cyanomelana* 白腹姬鶲 16–17cm

DESCRIPTION Large, long-winged flycatcher. Male has bright blue cap, dark blue neck and upperparts, and black face below eye. Chin, throat and upper breast black; remainder of underparts and sides of base of tail sparkling white. Female much duller than male, without blue in plumage and with no white at base of tail. Upperparts plain medium brown; underparts off-white with greyish cast to breast and flanks, and whitish, rather wide vertical throat-patch. First-winter male like female, but with blue wings, scapulars, lower back and tail. **SIMILAR SPECIES** See Hainan Blue Flycatcher (opposite). **HABITAT, HABITS AND STATUS** Passage migrant to wooded areas, locally common in spring, less frequent in autumn. (T, I)

Female   *Male*

Verditer Flycatcher
◾ *Eumyias thalassinus* 銅藍鶲 15–17cm

DESCRIPTION Rather large, longish-tailed, pale greenish-blue flycatcher with black lores. Female duller than male, with smaller and less obvious, sooty loral patch. **HABITAT, HABITS AND STATUS** Uncommon winter visitor to widespread woodland and wooded areas. Solitary. Conspicuous. Perches prominently, including on posts and wires in the open. (T, I)

▪ Shortwing & Robins ▪

Lesser Shortwing ▪ *Brachypteryx leucophris* 白喉短翅鶇 10–13cm

DESCRIPTION Small, robin-like bird with short wings and very short tail. Upperparts rufous-toned brown, with indistinct (often concealed) short white supercilium reaching just behind eye and narrow buffy eye-ring. Throat whitish; rest of underparts washed warm brown. Legs long and pale flesh in colour. Juvenile has heavily scaled brown and rufous-buff belly and vent, rufous-buff spotting on mantle and buff tips to greater coverts. **HABITAT, HABITS AND STATUS** Locally common and widespread resident, and winter visitor to woodland and thick shrubbery. Skulking, mainly on woodland floor. Favours damp upland forest. (T)

Rufous-tailed Robin ▪ *Larvivora sibilans* 紅尾歌鴝 13–14cm

DESCRIPTION Plain brown head, except for paler buffy loral stripe and eye-ring, and upperparts with conspicuously rufous uppertail-coverts and tail. Off-white underparts, copiously scaled olive-grey. Black bill and pink legs. Frequently cocks and pumps tail. **HABITAT, HABITS AND STATUS** Common winter visitor and passage migrant to widespread well-wooded areas throughout Hong Kong. Inconspicuous, quiet inhabitant of the forest floor. (T, I)

ROBINS

Bluethroat ■ *Luscinia svecica* 藍喉歌鴝 14cm

DESCRIPTION Grey-brown upperparts, bold, whitish supercilium and white crescent surrounding lower half of eye. Underparts pale buff-grey, and more buff on flanks. Adult breeding male displays bright blue chin, throat and upper breast, with red central spot and lower border of orange-red separated from blue breast by black and white band. Winter adult male has less solid colours on underparts than breeding male. Female and first-winter birds even less colourful, showing only traces of blue and orange tones or lacking them completely, but having black sides to throat with black breast-band mottled white and streaked black below. At all ages, rufous sides to base of tail conspicuous in flight. **HABITAT, HABITS AND STATUS** Skulking. Locally common winter visitor to agricultural and wet areas principally in north-west New Territories. (M, L)

Siberian Rubythroat
■ *Calliope calliope* 紅喉歌鴝 14cm

DESCRIPTION Upperparts mid-brown, rump and tail slightly richer. Greyish breast, more buff on flanks, and whitish belly and vent. Male has strikingly patterned head with bold white supercilium and submoustachial stripe, black malar and lores, and bright scarlet chin and throat. Female and first-winter birds have less well-defined submoustachial and malar stripes, and white chin and throat, though some individuals show pinkish-red blush. **HABITAT, HABITS AND STATUS** Common winter visitor and passage migrant to widespread areas of scrub and rough land throughout Hong Kong. Skulking and retiring ground dweller. (M, L, I)

Robins & Whistling Thrush

Red-flanked Bluetail ■ *Tarsiger cyanurus* 紅脇藍尾鴝 14cm

DESCRIPTION Male has bright blue head and upperparts, including rump and tail, pale supercilium and prominent reddish-orange flanks. Chin, throat and vent white; rest of underparts pale greyish-buff. Adult female has brown head and upperparts, white eye-ring and blue-washed tail. Triangular white bib from chin down through throat. Rest of underparts off-white with greyish-brown wash to breast. Orange flanks less extensive and not as bright as in male. First-winter birds similar to adult female. **HABITAT, HABITS AND STATUS** Common and widespread winter visitor and passage migrant to woods, parks and shrubland throughout Hong Kong. First-winter birds and adult females greatly predominate. (M, L, T, H, I)

Male

Female/first winter

Blue Whistling Thrush
■ *Myophonus caeruleus* 紫嘯鶇 29–35cm

DESCRIPTION Largest Hong Kong thrush. Full breasted and plump with strong black bill and legs. Deep blue overall, appearing black in the shade and glossy purplish-blue in sunlight. Head, mantle and covert-tips spotted paler violet. Voice: call a high-pitched, single whistle. **HABITAT, HABITS AND STATUS** Common and locally widespread resident along wooded streams, nullahs and rainwater channels throughout Hong Kong. Habitually fans tail when perched. (T, H, I)

OLD WORLD FLYCATCHERS

Yellow-rumped Flycatcher
■ *Ficedula zanthopygia* 白眉姬鶲 13–13.5cm

DESCRIPTION Small, strikingly bright black, yellow and white flycatcher. Male has black head and upperparts with short bold white supercilium and wing-patch. Underparts, lower back and rump vivid yellow. Vent and undertail-coverts white. Female much plainer than male, with grey-green upperparts, yellow rump, black tail, white wing-patch and black primaries. Underparts dingy white. **SIMILAR SPECIES** Vagrant male **Green-backed Flycatcher** *F. elisae* differs in having yellow supercilium, narrow eye-ring, vent and undertail-coverts, and greyish-green upperparts. Female duller than male, lacking yellow supercilium and rump, and white wing-patch. See also Narcissus Flycatcher (below). **HABITAT, HABITS AND STATUS** Uncommon autumn passage migrant, with rare spring records, to widespread wooded areas throughout Hong Kong. (M, T, I)

Narcissus Flycatcher
■ *Ficedula narcissina* 黃眉姬鶲 13–13.5cm

DESCRIPTION Male similar to Yellow-rumped Flycatcher (see above), but long, broad supercilium is yellow, and white wing-patch does not extend onto tertials. Female and first-winter birds plain and drab. Head and upperparts olive-brown, with brighter rufous-tinged rump and tail. Wings mid-brown. Underparts washed grey-brown with light mottling on sides of throat and breast. **SIMILAR SPECIES** See also Yellow-rumped Flycatcher, from which female and first-winter birds differ by lack of white wing-patch and no yellow on rump. **HABITAT, HABITS AND STATUS** Uncommon spring migrant with rare autumn records to widespread wooded areas. (M, T, I)

OLD WORLD FLYCATCHERS

Mugimaki Flycatcher ■ *Ficedula mugimaki* 鴝姬鶲 12.5–13.5cm

DESCRIPTION Small, brilliantly colourful black, white and orange flycatcher. Male's upperparts entirely black except for bold white flash behind eye, wing-patch and sides to base of tail. Throat and breast reddish-orange, and rest of underparts white. Female more subdued than male, with brown upperparts, and narrow white greater coverts bar and tertial fringes. Throat and breast pale orangey-buff, and rest of underparts white. First-winter male resembles adult female, but has more sooty upperparts, and suggestion of adult male's white head-flash and white sides to upper tail. **HABITAT, HABITS AND STATUS** Uncommon passage migrant mainly in autumn, scarce in spring and winter, to widespread woodland throughout Hong Kong. (T, I)

First-winter male

Taiga (Red-throated) Flycatcher ■ *Ficedula albicilla* 紅喉姬鶲 11.5cm

DESCRIPTION Small bird with greyish-brown upperparts, off-white underparts, pale eye-ring and white sides to base of black tail conspicuous in flight. Bill and legs blackish. Adult male in breeding plumage acquires red-orange chin and throat, the remnant of which is sometimes retained in winter. Voice: call a hard, trilled *trrrt*. **SIMILAR SPECIES** Scarce **Red-breasted Flycatcher** *F. parva* distinguishable at close range by pale base to lower mandible, and uppertail-coverts less extensively black than in Taiga. Call similar, but lower pitched and distinctly slower. **HABITAT, HABITS AND STATUS** Readily perches close to the ground and drops to feed. Commonly holds tail cocked. Common migrant and winter visitor to widespread, open woodland and shrubbery throughout Hong Kong. (M, L, I)

▪ REDSTARTS ▪

Daurian Redstart ▪ *Phoenicurus auroreus* 北紅尾鴝 15cm

DESCRIPTION Male splendidly bright with lavender-grey crown, nape and rear neck, and black forehead, 'face', throat, mantle, tail and wings, the last with conspicuous white patch. Underparts, rump and outer-tail feathers deep reddish-orange. First-winter male similar, but less solidly marked. Female has mid-brown upperparts with pale eye-ring, white wing-patch, buff underparts and reddish-orange rump and outer-tail feathers. **HABITAT, HABITS AND STATUS** Common winter visitor to widespread, open wooded and shrubby areas throughout Hong Kong. Habitually quivers tail when perched. (M, L, T, H, I)

Female

Male

Plumbeous Water Redstart
▪ *Phoenicurus fuliginosus* 紅尾水鴝 12–13cm

DESCRIPTION Male unmistakable. Head and body slate-blue, wings brownish-black, rump and tail bright orange-red. Female and first-winter male brownish-grey; blackish-brown wings with pale-tipped coverts produce two dotted wing-bars. Underparts grey, scaled white. Tail blackish-brown with conspicuous white uppertail, and patches at sides of base extending to tip.

HABITAT, HABITS AND STATUS Frequently wags and fans tail, perches on boulders and generally remains on the ground. Uncommon winter visitor to streams, water channels and reservoirs.

Male

Female

ROCK THRUSH & CHAT

Blue Rock Thrush ■ *Monticola solitarius* 藍磯鶇 20cm

DESCRIPTION Adult male has deep glossy blue head, breast, mantle and tail, black lores and wings, and deep orangey-red underparts. First-winter male similar to adult male though less bright, and scaly with red often almost lacking. Female brown, with dark blue wash above, and scaly light and dark brown 'face' and underparts. Subspecies *philippensis* described. Scarcer adult male *pandoo* has wholly blue underparts. **HABITAT, HABITS AND STATUS** Locally common passage migrant and winter visitor, and possibly scarce resident. Favours rocky coastal and upland sites throughout Hong Kong, though not averse to urban buildings. Tends to be solitary. (I)

Stejneger's Stonechat
■ *Saxicola stejnegeri* 黑喉石即鳥 12.5cm

DESCRIPTION Breeding male has black head, white collar and wing-patch, blackish upperparts, and black wings and tail, the last with white outer rectrices. Rump unstreaked white. Breast bright orange, and rest of underparts off-white. Winter male similarly patterned, but colours duller and far less contrasting; upperparts with pale brown fringes; rump unstreaked buff; underparts with orange-buff wash. Female and first-winter male browner, lacking dark hood, and orangey-buff washed breast. **HABITAT, HABITS AND STATUS** Common passage migrant and winter visitor to widespread open areas throughout Hong Kong. Perches conspicuously on twigs and bushes, from which it drops to the ground to catch insect food. (M, L, I)

▪ Leafbird & Flowerpeckers ▪

Orange-bellied Leafbird
▪ *Chloropsis hardwickii* 橙腹葉鵯 17–19.8cm

DESCRIPTION Breeding male has black head, white collar and wing-patch, blackish upperparts, and black wings and tail, the last with white outer rectrices. Rump unstreaked white. Breast bright orange, and rest of underparts off-white. Winter male similarly patterned, but its colours are duller and far less contrasting; upperparts with pale brown fringes; rump unstreaked buff; underparts with orange-buff wash. Female and first-winter male browner, lacking dark hood and with orange-buff washed breast. **HABITAT, HABITS AND STATUS** Uncommon resident and winter visitor to woodland of New Territories. Despite its gaudy plumage, it can be surprisingly difficult to locate, tending to favour the tree canopy in which it forages. (T)

Fire-breasted Flowerpecker ▪ *Dicaeum ignipectus* 紅胸啄花鳥 7–9cm

DESCRIPTION Tiny bird. Male has metallic greenish-blue head, upperparts, sides of breast and tail, and buff underparts with white throat, scarlet patch on breast and black line down centre of belly. Female has plain olive upperparts and buffy underparts. Voice: call a high-pitched, piercing *tsee*. **SIMILAR SPECIES** Scarce **Plain Flowerpecker** *D. minullum* closely resembles female Fire-breasted, but has paler 'face' and more olive underparts. **HABITAT, HABITS AND STATUS** Uncommon winter visitor and rare breeder in widespread woodland, particularly in Tai Po Kau. (T)

▪ FLOWERPECKERS & SUNBIRD ▪

Scarlet-backed Flowerpecker
▪ *Dicaeum cruentatum* 朱背啄花鳥 7–9cm

DESCRIPTION Tiny, bright and stub-tailed bird. Male has brilliant red cap, nape, mantle, rump and uppertail-coverts. 'Face', sides of neck and breast, wings and tail black with dark blue sheen. Underparts white; grey on flanks. Female plain olive-green above, with scarlet rump and uppertail-coverts, and black tail. Underparts off-white. Juvenile as female, but lacks scarlet plumage and has pink bill-base. Voice: call a hard, stony *chit*. **HABITAT, HABITS AND STATUS** Common resident, widespread in woodland and parkland mainly in New Territories. (M, L, T, H, I)

Male

Female

Fork-tailed Sunbird
▪ *Aethopyga christinae* 叉尾太陽鳥 10cm

DESCRIPTION Small bird. Male has metallic dark green crown and rear neck, black ear-covert patch surrounding eye and reaching base of bill, dark red chin, throat and breast, and rest of underparts pale yellow. Upperparts olive-green; rump yellow. Rather short, metallic dark green tail has white corner tips and elongated spiky central rectrices. Dark bill longish and strongly decurved. Female plain with dull green upperparts and pale yellow underparts. Tail has white corner tips but no elongated feathers. **HABITAT, HABITS AND STATUS** Common and widespread resident and winter visitor to wooded areas. Hong Kong's only resident sunbird. (M, L, T, H, I)

Female

Male

Sparrow & Munias

Eurasian Tree Sparrow
■ *Passer montanus* 樹麻雀 14cm

DESCRIPTION The common urban bird and the only sparrow in Hong Kong. Chestnut crown and nape, whitish cheeks and collar, black chin and spot at rear of ear-coverts. Upperparts warm brown, streaked black. White covert-tips form two well-defined wing-bars. Underparts sullied pale grey with buff flanks. **HABITAT, HABITS AND STATUS** Abundant and widespread resident in all types of residential and agricultural areas throughout Hong Kong. (M, L, T, H, I)

White-rumped Munia ■ *Lonchura striata* 白腰文鳥 11–12cm

DESCRIPTION Small, dark brown finch with heavy bill and longish, wedge-shaped tail. Brown head with black around bill-base and eye. Brown breast, mantle and rump, white lower back-patch, black tail and dull dark brown wings. Head and breast finely streaked and spotted buffy white. Underparts off-white. Large, cone-shaped bill grey with steel-blue sheen. Immature has paler brown upperparts and plainer buff underparts than adult. **HABITAT, HABITS AND STATUS** Common and widespread resident in open, wooded and shrubby areas and waste ground. Often forms large feeding flocks with Scaly-breasted Munia (see p. 148). (M, L, T, H, I)

MUNIAS, WAGTAILS & PIPITS

Scaly-breasted Munia
■ *Lonchura punctulata* 斑文鳥 12cm

DESCRIPTION Similar to White-rumped Munia (see p. 147) but head, upperparts including rump, and breast warmer brown. Underparts off-white, with feathers broadly edged dark brown to create scaly appearance. First-winter birds plain brown above with pale buffy-brown underparts. **HABITAT, HABITS AND STATUS** Abundant and widespread resident in open, grassy areas throughout Hong Kong. (M, L, T, H, I)

Forest Wagtail ■ *Dendronanthus indicus* 山鶺鴒 16–18cm

DESCRIPTION Unusually plumaged wagtail with grey-brown head and upperparts. Long white supercilium and striking wing pattern of black with bold yellow-white double wing-bars. Chin, throat and underparts off-white with black upper breast-band extending vertically down in centre between a lower second incomplete band. Long brown tail edged white. **HABITAT, HABITS AND STATUS** Uncommon passage migrant, less numerous in spring, with occasional records in winter. Widespread throughout Hong Kong in both open and closed wooded areas, especially Tai Po Kau. Inconspicuous, feeding quietly on the ground in the manner of pipits, and usually choosing to perch in a tree if flushed. Sways tail from side to side rather than up and down like other wagtails. (T)

WAGTAILS & PIPITS

Eastern Yellow Wagtail ■ *Motacilla tschutschensis* 東黃鶺鴒 17cm

DESCRIPTION Typical slim and long-tailed wagtail (though tail relatively shorter than tails of congeners). Three distinct subspecies occur in Hong Kong, the adults of which are easily separated in the field. Identification of immatures much more challenging. Females similar to males, but duller, and first-winter birds have plumage pattern of adult females, but are duller and greyer, with green and yellow tones reduced or even lacking. *M. t. tschutschensis* Adult breeding male has lavender-grey head; white long supercilium, crescent below eye and line on sides of throat. Chin and throat have variable amounts of white, from almost none to fairly extensive. Underparts bright yellow, often with band of dark streaking across breast of variable prominence and extent. Upperparts olive-green. Median and greater wing-coverts tipped white, producing narrow double wing-bars and tertials with broad white outer edges. Tail blackish with conspicuously white outer rectrices. Adult female and adult male winter birds similar but duller, with tendency to display more white on throat. First-winter birds sometimes have narrower and less prominent supercilium. *M. t. taivana* Adult breeding male differs in having rather dull brownish-green crown, ear-coverts, nape and mantle, normally darker ear-coverts and broad bright yellow supercilium. *M. t. macronyx* Adult breeding male differs in lacking supercilium; crown and nape deep slate-blue, and ear-coverts darker. **SIMILAR SPECIES** See Citrine Wagtail (see p. 150). **HABITAT, HABITS AND STATUS** Largely terrestrial, with high-stepping walk accompanied by head nodding and tail wagging. Runs swiftly. Only occasionally perches on low vegetation, and wags tail frequently. Favours wetlands, fish ponds, agricultural fields and grassy, open areas, mainly in northern New Territories. *Taivana* is by far the most numerous subspecies recorded in Hong Kong, being a common passage migrant and winter visitor. *Tschutschensis* is a common spring migrant and far less frequent in autumn, while *macronyx* is an uncommon passage migrant and winter visitor. (M, L, I)

M. t. tschutschensis

M. t. taivana

M. t. macronyx

■ Wagtails & Pipits ■

Citrine Wagtail ■ *Motacilla citreola* 黃頭鶺鴒 17cm

DESCRIPTION Superficially resembles Eastern Yellow Wagtail (see p.149). Adult winter male (full breeding plumage not usually attained in Hong Kong) has grey crown, ear-coverts and upperparts, black median and greater wing-coverts broadly tipped white, producing double wing-bars, and white-edged tertials. Tail black with conspicuous white outer rectrices. Yellow forehead, throat and underparts. Broad yellow supercilium curves around to encircle ear-coverts patch (some individuals lack ear-coverts patch and have wholly yellow 'face'). Adult winter female similar to male, with yellow on 'face', but head and upperparts browner and underparts off-white. First-winter birds similar to adult female winter, but initially lack all-yellow tones and are best identified by presence of pale surround to pale-centred, grey ear-coverts. **SIMILAR SPECIES** See Eastern Yellow Wagtail. **HABITAT, HABITS AND STATUS** Habitat and habits similar to Eastern Yellow Wagtail's. Uncommon winter visitor and passage migrant principally to Deep Bay area, particularly Long Valley. (L)

WAGTAILS & PIPITS

Grey Wagtail
■ *Motacilla cinerea* 灰鶺鴒 18–19cm

DESCRIPTION Soft grey head and mantle, yellow rump, white supercilium and malar, black wings and tail, the latter edged white. Lemon-yellow underparts, brighter towards rear. Adult male has black chin and throat in breeding plumage. **HABITAT, HABITS AND STATUS** Long tail constantly pumped. Flight deeply undulating. Common winter visitor and passage migrant with occasional summer records. Widespread along streams, reservoirs, fish ponds and marshes. (M, L, T, L, I)

White Wagtail ■ *Motacilla alba* 白鶺鴒 18cm

DESCRIPTION Distinctive black and white wagtail. Two subspecies regularly occur in Hong Kong, and another three are scarce or rare. M. *a. leucopsis* Adult breeding male has white head, throat, underparts and large wing-panel. Black rear crown, nape, mantle, rump, breast-patch and tail, the last with white outer rectrices. Winter male has smaller breast-patch. Female similar to male, but has duller and often greyer upperparts. First-winter birds have grey upperparts, crown and nape, variably grading from black to grey. M. *a. ocularis* Differs in having grey upperparts, black breast-patch more extensively covering chin and throat, and narrow black eye-stripe across white 'face'. Female similar to male but normally shows greyish centres to median and greater wing-coverts. First-winter birds similar to female, but with white wing-panel reduced to double white wing-bars.
SIMILAR SUBSPECIES Adult male of scarce M. *a. lugens* similar to *ocularis*, but mantle and back black not grey; rare *baicalensis* similar to *ocularis*, but lacks black eye-stripe, chin and throat; rare *personata* has black hood with white mask and grey upperparts.
HABITAT, HABITS AND STATUS Behaviour similar to Eastern Yellow Wagtail's (see p. 149). Widespread in Hong Kong, including in urban areas, with a preference for wetlands. Avoids woodland. *Leucopsis* is the most numerous subspecies, being a common resident; local population is heavily augmented by spring passage migrants and winter visitors. *Ocularis* is an uncommon passage migrant and winter visitor. (M, L, T, H, I)

M. a. leucopsis

M. a. ocularis

◾ PIPITS ◾

Richard's Pipit ◾ *Anthus richardi* 理氏鷚 18cm

DESCRIPTION Distinctly larger than all other regular lowland Hong Kong pipits. Medium-brown streaked crown and upperparts; ear-coverts plainer. Bold pale supercilium, and lores, chin and throat white. Narrow dark malar stripe broadens towards base to connect with streaking on neck and breast. Ground colour of underparts pale buff on breast merging to white elsewhere. Buff tips to median and greater coverts form double wing-bars, paler and more obvious in first-winter individuals. Voice: flight call loud, sharp *shrüp*. **HABITAT, HABITS AND STATUS** Long legs with long hind claws, and upright stance. Common passage migrant, winter visitor and local resident. Breeds on uplands of New Territories, while migrants visit grassy lowland areas. (M, L, I)

Olive-backed Pipit
◾ *Anthus hodgsoni* 樹鷚 14.5cm

DESCRIPTION Strong pale supercilium ending with distinct separate white spot and dark spot beneath at rear of ear-coverts. Crown and upperparts light brown with olive-green tone, finely streaked. Pale yellowish-buff throat with black malar. Breast and flanks off-white heavily streaked black. Belly white. Voice: call a high-pitched, thin *zeep*, often uttered when taking flight. **HABITAT, HABITS AND STATUS** Common and widespread winter visitor and passage migrant to open areas and woodland edges. Feeds on the ground in wooded areas and perches in trees when disturbed. Constantly pumps tail. (M, L, T, H, I)

▪ Pipits ▪

Red-throated Pipit ▪ *Anthus cervinus* 紅喉鷚 15cm

DESCRIPTION Upperparts brown heavily streaked black, with white mantle stripe and strongly streaked brown rump. Face, throat and upper breast brownish-orange of varying strength, though first-winter birds show no such warm blush and instead have pale, indistinct supercilium, uniform brown ear-coverts and black malar. Breast, flanks, belly and vent off-white; flanks with black streaks (extending across breast on first-winter birds). Voice: flight call a high-pitched *tsee*. **HABITAT, HABITS AND STATUS** Common passage migrant and winter visitor to marshland and open ground principally in Deep Bay area, but also regularly recorded elsewhere in New Territories. (M, L, I)

Buff-bellied Pipit ▪ *Anthus rubescens* 黃腹鷚 14–17cm

DESCRIPTION Rather dull pipit with lightly streaked crown, mantle and rump. Wide creamy supercilium merges into pale lores. Submoustachial, chin and throat creamy, separated by brown-grey malar. Whitish tips to coverts produce obvious double wing-bars. Underparts washed light buff with bold streaking on breast and flanks, the former concentrating as black patch at sides of throat. Bill blackish-grey with proximal section of lower mandible yellowish-brown. Legs pinkish- or brownish-yellow. Voice: call a rather shrill *tseep*. **HABITAT, HABITS AND STATUS** Uncommon passage migrant and winter visitor to wetlands and agricultural areas of northern New Territories, especially Long Valley. (M, L)

■ Pipits & Finch ■

Upland Pipit ■ *Anthus sylvanus* 山鷚 17cm

DESCRIPTION Large pipit recalling Richard's Pipit (see p. 152), from which it is distinguished by deeper toned, more heavily streaked upperparts and inconspicuous narrow black malar stripe. Underparts light buff, deeper on flanks, and lightly streaked black on breast and flanks. Bill rather short. Voice: call a high-pitched rapid, sustained *tit-tit-tit-tit-tit…*; song an un-pipit-like, far-carrying *weeeeech*. **HABITAT, HABITS AND STATUS** Uncommon resident restricted to grassy uplands of New Territories and Lantau. Usually sings from boulder. (I)

Chinese Grosbeak ■ *Eophona migratoria* 黑尾蠟嘴雀 15–18cm

DESCRIPTION Large, bulky finch with massive bill. Male has black head, and light grey-brown upperparts, paler on lower back and rump. Wings black glossed steel-blue, with small white patch at base of primaries, and prominent white tips to primaries and secondaries; black tail. Underparts greyish-buff with deeper orangey-buff breast and flank sides. Bill deep yellow with blackish tip and white border to base. Female duller than male and lacks black hood. **SIMILAR SPECIES Japanese Grosbeak** *E. personata*, a rare winter visitor, is larger with an even more massive bill. Black hood less extensive, reaching to just behind eye. Sexes are similar. Immature lacks hood but has black feathering around base of bill and on lores. **HABITAT, HABITS AND STATUS** Common and widespread winter visitor to woods and open areas with trees. Gregarious, typically wandering the countryside in small flocks. Also a scarce breeding species. (M, L, T, I)

▪ Buntings ▪

Tristram's Bunting
▪ *Emberiza tristrami* 白眉鵐 15cm

DESCRIPTION Rather small bunting. Male has bold black and white head pattern formed by white crown-stripe, supercilium and broad malar with inconspicuous white spot at rear of ear-coverts. Rest of head, chin and throat black. Dull brown rear neck and mantle, the latter strongly streaked black; richer, rufous-brown rump and tail with white outer rectrices. Breast and flanks warm buff; rest of underparts white. Female duller than male; black head stripes replaced with brown, chin and throat white. **HABITAT, HABITS AND STATUS** Uncommon winter visitor to widespread wooded areas, especially Tai Po Kau. Inconspicuous. (T, I)

Chestnut-eared Bunting ▪ *Emberiza fucata* 栗耳鵐 16cm

DESCRIPTION Male in breeding plumage has black-streaked grey crown and nape; chestnut ear-coverts with a small white spot at rear, and white malar, throat and breast decorated with black moustachial. Lateral throat-stripe and streaked upper breast-band, and thin rufous-chestnut lower breast-band. Mantle grey-brown streaked black, scapulars and rump rufous-chestnut, wings and tail brownish-black. Flanks buff streaked black, merging into white belly and vent. Winter male duller, developing wide supercilium, yellow at front grading to white at rear. Female and first-winter male similar, but less contrasting overall. **HABITAT, HABITS AND STATUS** Uncommon passage migrant, mainly in autumn, with occasional records in winter to open, grassy areas and thickets, predominantly in northern New Territories, especially Long Valley. (L)

First winter

■ Buntings ■

Little Bunting
■ *Emberiza pusilla* 小鵐 13–14cm

DESCRIPTION Small bunting with black-streaked brown upperparts, chestnut median crown-stripe, black lateral crown-stripe, broad buffy supercilium and malar, and distinctive chestnut lores and ear-coverts outlined in black. Moustachial black, and underparts white streaked black on breast and flanks. Narrow pale eye-ring, and white outer-tail feathers. Female and first-winter birds duller than adult male. HABITAT, HABITS AND STATUS Common winter visitor and passage migrant to widespread open areas throughout Hong Kong. (M, L, I)

Yellow-breasted Bunting ■ *Emberiza aureola* 黃胸鵐 14–15cm

DESCRIPTION Male in breeding plumage has dark rufous-brown crown, nape, upperparts and narrow breast-band, black 'face', and bright yellow underparts, including band below throat. White lesser coverts and broad tips to greater coverts create two bold wing-bars. Broad blackish streaks line flanks. Tail has white outer rectrices, which are conspicuous in flight. Winter male much duller, and female duller still, with broad pale supercilium, brown upperparts streaked black, and pale yellow underparts varying in strength. First-winter male similar to female, but head and underparts are washed pale lemon. HABITAT, HABITS AND STATUS Common and widespread autumn passage migrant, scarce in spring and rare in winter, to open, grassy areas and thickets, particularly Long Valley. (M, L)

■ BUNTINGS ■

Chestnut Bunting ■ *Emberiza rutila* 栗鵐 14–15cm

DESCRIPTION Male in breeding plumage an unmistakable combination of chestnut head, upperparts, throat and breast, and bright yellow underparts with some dark streaking on flanks. Winter male less bright with chestnut plumage fringed paler. Female duller. Head and upperparts grey-brown with dark streaking on mantle, chestnut lower back and rump only obvious in flight, tail brownish-black. Inconspicuous yellow-washed supercilium, narrow, pale eye-ring, yellowish-buff malar and underparts, dark throat-stripe and light streaking on flanks. First-winter male birds similar to female. **HABITAT, HABITS AND STATUS** Uncommon passage migrant, mainly in autumn, and occasional winter visitor to widespread open and shrubby areas. (L)

Female

Male

Black-faced Bunting ■ *Emberiza spodocephala* 栗耳鵐 13.5–15cm

DESCRIPTION Two subspecies occur. *E. s. sordida* male breeding plumage has greenish-olive hood and upper breast, black lores and chin, warm brown upperparts streaked blackish, white tips to lesser and greater coverts forming double wing-bar, bright yellow underparts with blackish streaking on flanks, and darkish brown tail with conspicuous white outer rectrices. Male in winter noticeably duller, with some individuals showing plumage pattern closer to female's. Female rather drab, mainly greyish-brown with weak supercilium and crown-stripe, indistinct dark streaking on breast and heavier streaking on flanks. Some individuals show weak yellow wash to underparts. First-winter birds similar to adult female. *E. s. spodocephala* resembles *sordida*, but is less bright. Hood and breast mid-grey and underparts pale lemon. **HABITAT, HABITS AND STATUS** Tends to remain in cover and often twitches tail. Common passage migrant and winter visitor (*sordida* far more numerous than *spodocephala*) to widespread open country throughout Hong Kong. As is the case with most other buntings, its population is showing recent marked decline. (M, L, I)

Female E. s. spodocephala

157

■ Checklist of the Birds of Hong Kong ■

Abbreviations

Abundance
A   Abundant
C   Common
E   Extinct in Hong Kong
R   Rare
S   Scarce
U   Uncommon
V   Vagrant

Hong Kong Status
Au   Autumn passage migrant
M   Passage migrant
Res   Resident
Sp   Spring passage migrant
Sum   Summer visitor
W   Winter visitor
-    Insufficient records to assess status
*   Species whose population is/was derived from captive stock

Where a species' Hong Kong status falls in two or more categories, those categories appear in the perceived order of importance. Thus, MW for Chestnut-eared Bunting indicates it is more of a passage migrant than a winter visitor whereas WM for Little Bunting is regarded as being more (or at least equally) a winter visitor than a passage migrant.

Global Status
World status of species follows BirdLife International's 'Threatened Birds of the World Birds to Watch 3'.
No listing is given for species of Least Concern; otherwise the categories are as follows:
NT   Near Threatened
VU   Vulnerable
EN   Endangered
CR   Critically Endangered

Common name	Chinese name	Abundance	Hong Kong Status	World Status
Anatidae Ducks, Geese & Swans				
Lesser Whistling Duck *Dendrocygna javanica*	栗樹鴨	R	MSum	
Taiga Bean Goose *Anser fabalis*	寒林豆雁	V	-	
Tundra Bean Goose *Anser serrirostris*	凍原豆雁	V	-	
Greylag Goose *Anser anser*	灰雁	V	-	
Greater White-fronted Goose *Anser albifrons*	白額雁	V	-	
Lesser White-fronted Goose *Anser erythropus*	小白額雁	V	-	VU
Whooper Swan *Cygnus cygnus*	大天鵝	V	-	
Common Shelduck *Tadorna tadorna*	翹鼻麻鴨	S	W	
Ruddy Shelduck *Tadorna ferruginea*	赤麻鴨	R	W	
Mandarin Duck *Aix galericulata*	鴛鴦	R	W	
Cotton Pygmy Goose *Nettapus coromandelianus*	棉鳧	V	-	
Gadwall *Anas strepera*	赤膀鴨	U	W	
Falcated Duck *Anas falcata*	羅紋鴨	U	W	NT
Eurasian Wigeon *Anas penelope*	赤頸鴨	A	W	

Common name	Chinese name	Abundance	Hong Kong Status	World Status
American Wigeon *Anas americana*	綠眉鴨	V		
Mallard *Anas platyrhynchos*	綠頭鴨	S	W	
Philippine Duck *Anas luzonica*	棕頸鴨	V	-	
Indian Spot-billed Duck *Anas poecilorhyncha*	印緬斑嘴鴨	R	Sum	
Chinese Spot-billed Duck *Anas zonorhyncha*	中華斑嘴鴨	U	W	
Northern Shoveler *Anas clypeata*	琵嘴鴨	A	W	
Northern Pintail *Anas acuta*	針尾鴨	A	W	
Garganey *Anas querquedula*	白眉鴨	C	MW	
Baikal Teal *Anas formosa*	花臉鴨	R	W	
Eurasian Teal *Anas crecca*	綠翅鴨	A	W	
Green-winged Teal *Anas carolinensis*	美洲綠翅鴨	V	-	
Red-crested Pochard *Netta rufina*	赤嘴潛鴨	V	-	
Common Pochard *Aythya ferina*	紅頭潛鴨	S	W	VU
Baer's Pochard *Aythya baeri*	青頭潛鴨	R	W	CR
Ferruginous Duck *Aythya nyroca*	白眼潛鴨	R	W	NT
Tufted Duck *Aythya fuligula*	鳳頭潛鴨	A	W	
Greater Scaup *Aythya marila*	斑背潛鴨	S	W	
White-winged Scoter *Melanitta deglandi*	斑臉海番鴨	V	-	
Black Scoter *Melanitta americana*	黑海番鴨	V	-	NT
Common Goldeneye *Bucephala clangula*	鵲鴨	V	-	
Smew *Mergellus albellus*	白秋沙鴨	V	-	
Red-breasted Merganser *Mergus serrator*	紅胸秋沙鴨	S	Sp	
Phasianidae Pheasants and allies				
Chinese Francolin *Francolinus pintadeanus*	中華鷓鴣	C	Res	
Japanese Quail *Coturnix japonica*	鵪鶉	U	MW	NT
Gaviidae Loons				
Red-throated Loon *Gavia stellata*	紅喉潛鳥	V	-	
Pacific Loon *Gavia pacifica*	太平洋潛鳥	V	-	
Yellow-billed Loon *Gavia adamsii*	黃嘴潛鳥	V	-	NT
Procellariidae Petrels, Shearwaters				
Streaked Shearwater *Calonectris leucomelas*	白額鸌	S	Sp	NT
Wedge-tailed Shearwater *Puffinus pacificus*	楔尾鸌	V	-	
Short-tailed Shearwater *Puffinus tenuirostris*	短尾鸌	U	Sp	
Bulwer's Petrel *Bulweria bulwerii*	褐燕鸌	V	-	
Podicipedidae Grebes				
Little Grebe *Tachybaptus ruficollis*	小鸊鷉	C	Res	
Great Crested Grebe *Podiceps cristatus*	鳳頭鸊鷉	C	W	
Horned Grebe *Podiceps auritus*	角鸊鷉	V	-	VU
Black-necked Grebe *Podiceps nigricollis*	黑頸鸊鷉	V	-	
Phaethontidae Tropicbirds				
White-tailed Tropicbird *Phaethon lepturus*	白尾鸏	V	-	
Ciconiidae Storks				
Black Stork *Ciconia nigra*	黑鸛	R	W	
Oriental Stork *Ciconia boyciana*	東方白鸛	R	W	EN
Threskiornithidae Ibises, Spoonbills				
Black-headed Ibis *Threskiornis melanocephalus*	黑頭白䴉	R	W	NT
Glossy Ibis *Plegadis falcinellus*	彩䴉	V	-	
Eurasian Spoonbill *Platalea leucorodia*	白琵鷺	U	W	
Black-faced Spoonbill *Platalea minor*	黑臉琵鷺	C	WSum	EN

Common name	Chinese name	Abundance	Hong Kong Status	World Status
Ardeidae Herons, Bitterns				
Eurasian Bittern *Botaurus stellaris*	大麻鳽	U	WSp	
Yellow Bittern *Ixobrychus sinensis*	黃葦鳽	U	MSumW	
Von Schrenck's Bittern *Ixobrychus eurhythmus*	紫背葦鳽	S	M	
Cinnamon Bittern *Ixobrychus cinnamomeus*	栗葦鳽	U	MWSum	
Black Bittern *Dupetor flavicollis*	黑鳽	S	M	
Japanese Night Heron *Gorsachius goisagi*	栗鳽	V	-	EN
Malayan Night Heron *Gorsachius melanolophus*	黑冠鳽	R	SpSum	
Black-crowned Night Heron *Nycticorax nycticorax*	夜鷺	C	ResM	
Striated Heron *Butorides striata*	綠鷺	C	SumMW	
Chinese Pond Heron *Ardeola bacchus*	池鷺	C	SumMW	
Eastern Cattle Egret *Bubulcus coromandus*	牛背鷺	C	SumMW	
Grey Heron *Ardea cinerea*	蒼鷺	C	WSum	
Purple Heron *Ardea purpurea*	草鷺	U	MWSum	
Great Egret *Ardea alba*	大白鷺	A	WSumM	
Intermediate Egret *Egretta intermedia*	中白鷺	C	MWSum	
Little Egret *Egretta garzetta*	小白鷺	A	ResMW	
Pacific Reef Heron *Egretta sacra*	岩鷺	C	Res	
Swinhoe's Egret *Egretta eulophotes*	黃嘴白鷺	S	Sp	VU
Pelecanidae Pelicans				
Dalmatian Pelican *Pelecanus crispus*	卷羽鵜鶘	R	W	VU
Fregatidae Frigatebirds				
Christmas Island Frigatebird *Fregata andrewsi*	白腹軍艦鳥	V	-	CR
Great Frigatebird *Fregata minor*	黑腹軍艦鳥	V	-	
Lesser Frigatebird *Fregata ariel*	白斑軍艦鳥	S	SpSum	
Sulidae Gannets, Boobies				
Masked Booby *Sula dactylatra*	藍臉鰹鳥	V	-	
Red-footed Booby *Sula sula*	紅腳鰹鳥	V	-	
Brown Booby *Sula leucogaster*	褐鰹鳥	V	-	
Phalacrocoracidae Cormorants, Shags				
Great Cormorant *Phalacrocorax carbo*	普通鸕鶿	A	W	
Japanese Cormorant *Phalacrocorax capillatus*	綠背鸕鶿	V	-	
Pandionidae Ospreys				
Western Osprey *Pandion haliaetus*	鶚	C	W	
Accipitridae Kites, Hawks and Eagles				
Black-winged Kite *Elanus caeruleus*	黑翅鳶	U	MWSum	
Crested Honey Buzzard *Pernis ptilorhyncus*	鳳頭蜂鷹	U	AuW	
Black Baza *Aviceda leuphotes*	黑冠鵑隼	S	MSum	
Eurasian Black Vulture *Aegypius monachus*	禿鷲	R	W	NT
Crested Serpent Eagle *Spilornis cheela*	蛇鵰	C	Res	
Mountain Hawk Eagle *Nisaetus nipalensis*	鷹鵰	V	-	
Greater Spotted Eagle *Clanga clanga*	烏鵰	C	W	VU
Steppe Eagle *Aquila nipalensis*	草原鵰	V	-	EN
Eastern Imperial Eagle *Aquila heliaca*	白肩鵰	C	W	VU
Bonelli's Eagle *Aquila fasciata*	白腹隼	U	Res	
Crested Goshawk *Accipiter trivirgatus*	鳳頭鷹	C	Res	
Chinese Sparrowhawk *Accipiter soloensis*	赤腹鷹	C	M	
Japanese Sparrowhawk *Accipiter gularis*	日本松雀鷹	U	MW	

Common name	Chinese name	Abundance	Hong Kong Status	World Status
Besra *Accipiter virgatus*	松雀鷹	C	ResM	
Eurasian Sparrowhawk *Accipiter nisus*	雀鷹	S	MW	
Northern Goshawk *Accipiter gentilis*	蒼鷹	V	-	
Eastern Marsh Harrier *Circus spilonotus*	白腹鷂	C	W	
Pied Harrier *Circus melanoleucos*	鵲鷂	U	MW	
Black Kite *Milvus migrans*	黑鳶	A	WRes	
Brahminy Kite *Haliastur indus*	栗鳶	V	-	
White-bellied Sea Eagle *Haliaeetus leucogaster*	白腹海鵰	C	Res	
Grey-faced Buzzard *Butastur indicus*	灰臉鵟鷹	U	M	
Eastern Buzzard *Buteo japonicus*	普通鵟	C	W	
Rallidae Rails, Crakes and Coots				
Slaty-legged Crake *Rallina eurizonoides*	灰腳秧雞	C	SumMW	
Slaty-breasted Crake *Gallirallus striatus*	灰胸秧雞	S	ResM	
Western Water Rail *Rallus aquaticus*	西方秧雞	V	-	
Eastern Water Rail *Rallus indicus*	普通秧雞	S	MW	
Brown Crake *Amaurornis akool*	紅腳苦惡鳥	R	MSum	
White-breasted Waterhen *Amaurornis phoenicurus*	白胸苦惡鳥	C	Res	
Baillon's Crake *Porzana pusilla*	小田雞	S	M	
Ruddy-breasted Crake *Porzana fusca*	紅胸田雞	U	MW	
Band-bellied Crake *Porzana paykullii*	斑脇田雞	V	-	NT
White-browed Crake *Porzana cinerea*	白眉田雞	V	-	
Watercock *Gallicrex cinerea*	董雞	S	M	
Black-backed Swamphen *Porphyrio indicus*	黑背紫水雞	V	-	
Common Moorhen *Gallinula chloropus*	黑水雞	C	WResM	
Eurasian Coot *Fulica atra*	骨頂雞	U	W	
Gruidae Cranes				
Siberian Crane *Grus leucogeranus*	白鶴	V	-	CR
Common Crane *Grus grus*	灰鶴	V	-	
Turnicidae Button-quail				
Yellow-legged Button-quail *Turnix tanki*	黃腳三趾鶉	S	AuW	
Barred Button-quail *Turnix suscitator*	棕三趾鶉	R	AuW	
Burhinidae Stone-curlews, Thick-knees				
Great Stone-curlew *Esacus recurvirostris*	大石鴴	V	-	NT
Haematopodidae Oystercatchers				
Eurasian Oystercatcher *Haematopus ostralegus*	蠣鷸	V	-	NT
Recurvirostridae Stilts, Avocets				
Black-winged Stilt *Himantopus himantopus*	黑翅長腳鷸	C	MWSum	
Pied Avocet *Recurvirostra avosetta*	反嘴鷸	A	W	
Charadriidae Plovers				
Northern Lapwing *Vanellus vanellus*	鳳頭麥雞	S	W	NT
Grey-headed Lapwing *Vanellus cinereus*	灰頭麥雞	U	WM	
European Golden Plover *Pluvialis apricaria*	歐金鴴	V	-	
Pacific Golden Plover *Pluvialis fulva*	太平洋金斑鴴	C	WM	
Grey Plover *Pluvialis squatarola*	灰斑鴴	A	WP	
Common Ringed Plover *Charadrius hiaticula*	劍鴴	R	W	
Long-billed Plover *Charadrius placidus*	長嘴鴴	V	-	
Little Ringed Plover *Charadrius dubius*	金眶鴴	C	WMSum	
Kentish Plover *Charadrius alexandrinus*	環頸鴴	A	WM	

Common name	Chinese name	Abundance	Hong Kong Status	World Status
Lesser Sand Plover *Charadrius mongolus*	蒙古沙鴴	U	MW	
Greater Sand Plover *Charadrius leschenaultii*	鐵嘴沙鴴	A	MW	
Oriental Plover *Charadrius veredus*	東方鴴	S	M	
Rostratulidae Painted-snipes				
Greater Painted-snipe *Rostratula benghalensis*	彩鷸	U	Res	
Jacanidae Jacanas				
Pheasant-tailed Jacana *Hydrophasianus chirurgus*	水雉	U	M	
Scolopacidae Sandpipers, Snipes				
Eurasian Woodcock *Scolopax rusticola*	丘鷸	U	AuW	
Pintail Snipe *Gallinago stenura*	針尾沙錐	C	MW	
Swinhoe's Snipe *Gallinago megala*	大沙錐	S	MW	
Common Snipe *Gallinago gallinago*	扇尾沙錐	C	WM	
Long-billed Dowitcher *Limnodromus scolopaceus*	長嘴鷸	S	M	
Asian Dowitcher *Limnodromus semipalmatus*	半蹼鷸	C	M	NT
Black-tailed Godwit *Limosa limosa*	黑尾塍鷸	A	MW	NT
Bar-tailed Godwit *Limosa lapponica*	斑尾塍鷸	U	MW	NT
Little Curlew *Numenius minutus*	小杓鷸	R	M	
Whimbrel *Numenius phaeopus*	中杓鷸	C	MW	
Eurasian Curlew *Numenius arquata*	白腰杓鷸	A	W	NT
Far Eastern Curlew *Numenius madagascariensis*	紅腰杓鷸	U	MW	EN
Spotted Redshank *Tringa erythropus*	鶴鷸	C	MW	
Common Redshank *Tringa totanus*	紅腳鷸	A	MW	
Marsh Sandpiper *Tringa stagnatilis*	澤鷸	A	WM	
Common Greenshank *Tringa nebularia*	青腳鷸	A	WM	
Nordmann's Greenshank *Tringa guttifer*	小青腳鷸	U	MW	EN
Lesser Yellowlegs *Tringa flavipes*	小黃腳鷸	V	-	
Green Sandpiper *Tringa ochropus*	白腰草鷸	C	MW	
Wood Sandpiper *Tringa glareola*	林鷸	C	MW	
Grey-tailed Tattler *Tringa brevipes*	灰尾漂鷸	C	M	NT
Terek Sandpiper *Xenus cinereus*	翹嘴鷸	C	MW	
Common Sandpiper *Actitis hypoleucos*	磯鷸	C	MWSum	
Ruddy Turnstone *Arenaria interpres*	翻石鷸	C	MW	
Great Knot *Calidris tenuirostris*	大濱鷸	C	MW	EN
Red Knot *Calidris canutus*	紅腹濱鷸	C	MW	NT
Sanderling *Calidris alba*	三趾濱鷸	U	M	
Red-necked Stint *Calidris ruficollis*	紅頸濱鷸	A	MW	NT
Little Stint *Calidris minuta*	小濱鷸	U	Sp	
Temminck's Stint *Calidris temminckii*	青腳濱鷸	C	WM	
Long-toed Stint *Calidris subminuta*	長趾濱鷸	C	MW	
Pectoral Sandpiper *Calidris melanotos*	斑胸濱鷸	R	M	
Sharp-tailed Sandpiper *Calidris acuminata*	尖尾濱鷸	C	M	
Curlew Sandpiper *Calidris ferruginea*	彎嘴濱鷸	A	MW	NT
Dunlin *Calidris alpina*	黑腹濱鷸	A	WM	
Spoon-billed Sandpiper *Eurynorhynchus pygmeus*	勺嘴鷸	S	MW	CR
Broad-billed Sandpiper *Limicola falcinellus*	闊嘴鷸	C	MW	
Buff-breasted Sandpiper *Tryngites subruficollis*	飾胸鷸	V	-	NT
Ruff *Philomachus pugnax*	流蘇鷸	S	MW	
Red-necked Phalarope *Phalaropus lobatus*	紅頸瓣蹼鷸	C	MW	

Common name	Chinese name	Abundance	Hong Kong Status	World Status
Red Phalarope *Phalaropus fulicarius*	灰瓣蹼鷸	V	-	
Glareolidae Pratincoles, Coursers				
Oriental Pratincole *Glareola maldivarum*	普通燕鴴	C	M	
Small Pratincole *Glareola lacteal*	灰燕鴴	V	-	
Laridae Gulls, Terns and Skimmers				
Brown Noddy *Anous stolidus*	白頂玄鷗	V	-	
Black-legged Kittiwake *Rissa tridactyla*	三趾鷗	R	SpW	
Slender-billed Gull *Chroicocephalus genei*	細嘴鷗	V	-	
Brown-headed Gull *Chroicocephalus brunnicephalus*	棕頭鷗	R	WM	
Black-headed Gull *Chroicocephalus ridibundus*	紅嘴鷗	A	W	
Saunders's Gull *Chroicocephalus saundersi*	黑嘴鷗	C	W	VU
Little Gull *Hydrocoloeus minutus*	小鷗	V	-	
Franklin's Gull *Leucophaeus pipixcan*	弗氏鷗	V	-	
Relict Gull *Ichthyaetus relictus*	遺鷗	V	-	VU
Pallas's Gull *Ichthyaetus ichthyaetus*	漁鷗	S	WSp	
Black-tailed Gull *Larus crassirostris*	黑尾鷗	C	WM	
Mew Gull *Larus canus*	海鷗	S	WSM	
Glaucous-winged Gull *Larus glaucescens*	灰翅鷗	V	W	
Glaucous Gull *Larus hyperboreus*	北極鷗	R	W	
Vega Gull *Larus vegae*	織女銀鷗	S	W	
Caspian Gull *Larus cachinnans*	蒙古銀鷗	U	W	
Slaty-backed Gull *Larus schistisagus*	灰背鷗	S	W	
Heuglin's Gull *Larus fuscus*	烏灰銀鷗	C	WM	
Gull-billed Tern *Gelochelidon nilotica*	鷗嘴噪鷗	C	M	
Caspian Tern *Hydroprogne caspia*	紅嘴巨鷗	C	MW	
Greater Crested Tern *Thalasseus bergii*	大鳳頭燕鷗	C	M	
Little Tern *Sternula albifrons*	白額燕鷗	U	M	
Aleutian Tern *Onychoprion aleuticus*	白腰燕鷗	U	M	
Bridled Tern *Onychoprion anaethetus*	褐翅燕鷗	C	SumM	
Sooty Tern *Onychoprion fuscatus*	烏燕鷗	V	-	
Roseate Tern *Sterna dougallii*	粉紅燕鷗	U	Sum	
Black-naped Tern *Sterna sumatrana*	黑枕燕鷗	C	SumM	
Common Tern *Sterna hirundo*	普通燕鷗	U	M	
Whiskered Tern *Chlidonias hybrida*	鬚浮鷗	C	MW	
White-winged Tern *Chlidonias leucopterus*	白翅浮鷗	C	M	
Stercorariidae Skuas				
Pomarine Skua *Stercorarius pomarinus*	中賊鷗	S	M	
Parasitic Jaeger *Stercorarius parasiticus*	短尾賊鷗	S	Sp	
Long-tailed Jaeger *Stercorarius longicaudus*	長尾賊鷗	U	M	
Alcidae Auks				
Ancient Murrelet *Synthliboramphus antiquus*	扁嘴海雀	U	SpW	
Japanese Murrelet *Synthliboramphus wumizusume*	冠海雀	V	-	VU
Columbidae Pigeons, Doves				
Domestic Pigeon *Columba livia*	原鴿	C	Res*	
Oriental Turtle Dove *Streptopelia orientalis*	山斑鳩	C	W	
Eurasian Collared Dove *Streptopelia decaocto*	灰斑鳩	U	Res*	
Red Turtle Dove *Streptopelia tranquebarica*	火斑鳩	C	MW	
Spotted Dove *Spilopelia chinensis*	珠頸斑鳩	A	Res	

Common name	Chinese name	Abundance	Hong Kong Status	World Status
Barred Cuckoo Dove *Macropygia unchall*	斑尾鵑鳩	V	-	
Common Emerald Dove *Chalcophaps indica*	綠翅金鳩	U	Res	
Orange-breasted Green Pigeon *Treron bicinctus*	橙胸綠鳩	V	-	
Wedge-tailed Green Pigeon *Treron sphenurus*	楔尾綠鳩	V	-	
Thick-billed Green Pigeon *Treron curvirostra*	厚嘴綠鳩	V	-	
White-bellied Green Pigeon *Treron sieboldii*	紅翅綠鳩	V	-	
Whistling Green Pigeon *Treron formosae*	紅頂綠鳩	V	-	
Cuculidae Cuckoos				
Greater Coucal *Centropus sinensis*	褐翅鴉鵑	C	Res	
Lesser Coucal *Centropus bengalensis*	小鴉鵑	U	Res	
Chestnut-winged Cuckoo *Clamator coromandus*	紅翅鳳頭鵑	U	SumM	
Asian Koel *Eudynamys scolopaceus*	噪鵑	C	SumW	
Plaintive Cuckoo *Cacomantis merulinus*	八聲杜鵑	C	SumW	
Fork-tailed Drongo Cuckoo *Surniculus dicruroides*	烏鵑	R	M	
Large Hawk Cuckoo *Hierococcyx sparverioides*	大鷹鵑	C	SpSum	
Northern Hawk Cuckoo *Hierococcyx hyperythrus*	北鷹鵑	V	-	
Hodgson's Hawk Cuckoo *Hierococcyx nisicolor*	霍氏鷹鵑	U	SpSum	
Lesser Cuckoo *Cuculus poliocephalus*	小杜鵑	R	M	
Indian Cuckoo *Cuculus micropterus*	四聲杜鵑	C	SpSum	
Oriental Cuckoo *Cuculus optatus*	東方中杜鵑	S	M	
Common Cuckoo *Cuculus canorus*	大杜鵑	V	-	
Tytonidae Barn Owls				
Eastern Grass Owl *Tyto longimembris*	草鴞	V	-	
Strigidae True Owls				
Collared Scops Owl *Otus lettia*	領角鴞	C	Res	
Oriental Scops Owl *Otus sunia*	紅角鴞	S	Au	
Eurasian Eagle Owl *Bubo bubo*	鵰鴞	S	Res	
Brown Fish Owl *Ketupa zeylonensis*	褐漁鴞	S	Res	
Brown Wood Owl *Strix leptogrammica*	褐林鴞	S	Res	
Asian Barred Owlet *Glaucidium cuculoides*	斑頭鵂鶹	C	Res	
Northern Boobook *Ninox japonica*	鷹鴞	U	M	
Short-eared Owl *Asio flammeus*	短耳鴞	V	-	
Caprimulgidae Nightjars				
Grey Nightjar *Caprimulgus jotaka*	普通夜鷹	S	MSum	
Savanna Nightjar *Caprimulgus affinis*	林夜鷹	U	Res	
Apodidae Swifts				
Himalayan Swiftlet *Aerodramus brevirostris*	短嘴金絲燕	S	MW	
White-throated Needletail *Hirundapus caudacutus*	白喉針尾雨燕	S	M	
Silver-backed Needletail *Hirundapus cochinchinensis*	灰喉針尾雨燕	S	M	
Brown-backed Needletail *Hirundapus giganteus*	褐背針尾雨燕	V	-	
Common Swift *Apus apus*	普通雨燕	V	-	
Pacific Swift *Apus pacificus*	白腰雨燕	C	MSumW	
House Swift *Apus nipalensis*	小白腰雨燕	A	SpRes	
Coraciidae Rollers				
Eurasian Roller *Coracias garrulus*	藍胸佛法僧	V	-	
Oriental Dollarbird *Eurystomus orientalis*	三寶鳥	C	M	
Alcedinidae Kingfishers				
Ruddy Kingfisher *Halcyon coromanda*	赤翡翠	V	-	

Common name	Chinese name	Abundance	Hong Kong Status	World Status
White-throated Kingfisher *Halcyon smyrnensis*	白胸翡翠	C	Res	
Black-capped Kingfisher *Halcyon pileata*	藍翡翠	U	MW	
Collared Kingfisher *Todiramphus chloris*	白領翡翠	V	-	
Common Kingfisher *Alcedo atthis*	普通翠鳥	C	MWSum	
Oriental Dwarf Kingfisher *Ceyx erithaca*	三趾翠鳥	V	-	
Crested Kingfisher *Megaceryle lugubris*	冠魚狗	R	Res	
Pied Kingfisher *Ceryle rudis*	斑魚狗	C	Res	
Meropidae Bee-eaters				
Blue-tailed Bee-eater *Merops philippinus*	栗喉蜂虎	U	M	
Blue-throated Bee-eater *Merops viridis*	藍喉蜂虎	V	-	
Upupidae Hoopoes				
Eurasian Hoopoe *Upupa epops*	戴勝	U	WM	
Megalaimidae Asian Barbets				
Great Barbet *Psilopogon virens*	大擬啄木鳥	U	Res	
Chinese Barbet *Psilopogon faber*	黑眉擬啄木鳥	V	-	
Picidae Woodpeckers				
Eurasian Wryneck *Jynx torquilla*	蟻鴷	U	MW	
Speckled Piculet *Picumnus innominatus*	斑姬啄木鳥	R	Res	
Rufous-bellied Woodpecker *Dendrocopos hyperythrus*	棕腹啄木鳥	V	-	
Grey-headed Woodpecker *Picus canus*	灰頭綠啄木鳥	V	-	
Bay Woodpecker *Blythipicus pyrrhotis*	黃嘴栗啄木鳥	R	Res	
Rufous Woodpecker *Micropternus brachyurus*	栗啄木鳥	V	-	
Falconidae Caracaras, Falcons				
Common Kestrel *Falco tinnunculus*	紅隼	C	AuW	
Amur Falcon *Falco amurensis*	阿穆爾隼	U	Au	
Eurasian Hobby *Falco subbuteo*	燕隼	U	MSum	
Peregrine Falcon *Falco peregrinus*	遊隼	C	ResW	
Cacatuidae Cockatoos				
Yellow-crested Cockatoo *Cacatua sulphurea*	小葵花鳳頭鸚鵡	C	Res*	CR
Psittaculidae Old World Parrots				
Alexandrine Parakeet *Psittacula eupatria*	亞歷山大鸚鵡	U	Res*	NT
Rose-ringed Parakeet *Psittacula krameri*	紅領綠鸚鵡	C	Res*	
Pittidae Pittas				
Fairy Pitta *Pitta nympha*	仙八色鶇	R	M	VU
Blue-winged Pitta *Pitta moluccensis*	藍翅八色鶇	V	-	
Artamidae Woodswallows, Butcherbirds and allies				
Ashy Woodswallow *Artamus fuscus*	灰燕鵙	V	-	
Campephagidae Cuckooshrikes				
Black-winged Cuckooshrike *Coracina melaschistos*	暗灰鵑鵙	C	MW	
Rosy Minivet *Pericrocotus roseus*	粉紅山椒鳥	V	-	
Swinhoe's Minivet *Pericrocotus cantonensis*	小灰山椒鳥	S	M	
Ashy Minivet *Pericrocotus divaricatus*	灰山椒鳥	U	M	
Grey-chinned Minivet *Pericrocotus solaris*	灰喉山椒鳥	C	Res	
Scarlet Minivet *Pericrocotus speciosus*	赤紅山椒鳥	C	Res	
Laniidae Shrikes				
Tiger Shrike *Lanius tigrinus*	虎紋伯勞	R	Au	
Bull-headed Shrike *Lanius bucephalus*	牛頭伯勞	S	AuW	
Brown Shrike *Lanius cristatus*	紅尾伯勞	C	MW	

Common name	Chinese name	Abundance	Hong Kong Status	World Status
Red-backed Shrike *Lanius collurio*	紅背伯勞	V	-	
Long-tailed Shrike *Lanius schach*	棕背伯勞	C	Res	
Grey-backed Shrike *Lanius tephronotus*	灰背伯勞	V	-	
Vireonidae Vireos, Greenlets				
White-bellied Erpornis *Erpornis zantholeuca*	白腹鳳鶥	U	Res	
Oriolidae Figbirds, Orioles				
Black-naped Oriole *Oriolus chinensis*	黑枕黃鸝	C	MW	
Dicruridae Drongos				
Black Drongo *Dicrurus macrocercus*	黑卷尾	C	MSumW	
Ashy Drongo *Dicrurus leucophaeus*	灰卷尾	U	W	
Crow-billed Drongo *Dicrurus annectans*	鴉嘴卷尾	V	-	
Hair-crested Drongo *Dicrurus hottentottus*	髮冠卷尾	C	WMRes	
Monarchidae Monarchs				
Black-naped Monarch *Hypothymis azurea*	黑枕王鶲	U	WM	
Amur Paradise-Flycatcher *Terpsiphone incei*	綬帶	U	MW	
Japanese Paradise-Flycatcher *Terpsiphone atrocaudata*	紫綬帶	U	M	NT
Corvidae Crows, Jays				
Eurasian Jay *Garrulus glandarius*	松鴉	R	Res	
Azure-winged Magpie *Cyanopica cyanus*	灰喜鵲	C	Res	
Red-billed Blue Magpie *Urocissa erythrorhyncha*	紅嘴藍鵲	C	Res	
Grey Treepie *Dendrocitta formosae*	灰樹鵲	U	Res	
Eurasian Magpie *Pica pica*	喜鵲	C	Res	
Daurian Jackdaw *Coloeus dauuricus*	達烏里寒鴉	R	W	
House Crow *Corvus splendens*	家鴉	C	Res*	
Carrion Crow *Corvus corone*	小嘴烏鴉	R	W	
Collared Crow *Corvus torquatus*	白頸鴉	C	Res	NT
Large-billed Crow *Corvus macrorhynchos*	大嘴烏鴉	C	Res	
Bombycillidae Waxwings				
Japanese Waxwing *Bombycilla japonica*	小太平鳥	V	-	NT
Stenostiridae Fairy Flycatchers				
Grey-headed Canary-flycatcher *Culicicapa ceylonensis*	方尾鶲	U	W	
Paridae Tits, Chickadees				
Yellow-bellied Tit *Pardaliparus venustulus*	黃腹山雀	R	W	
Varied Tit *Sittiparus varius*	雜色山雀	V	-	
Japanese Tit *Parus minor*	日本山雀	V	-	
Cinereous Tit *Parus cinereus*	蒼背山雀	C	Res	
Yellow-cheeked Tit *Machlolophus spilonotus*	黃頰山雀	U	Res*	
Remizidae Penduline Tits				
Chinese Penduline Tit *Remiz consobrinus*	中華攀雀	C	MW	
Alaudidae Larks				
Oriental Skylark *Alauda gulgula*	小雲雀	S	AuW	
Eurasian Skylark *Alauda arvensis*	雲雀	U	AuW	
Mongolian Short-toed Lark *Calandrella dukhunensis*	蒙古短趾百靈	V	-	
Asian Short-toed Lark *Alaudala cheleensis*	亞洲短趾百靈	V	-	
Pycnonotidae Bulbuls				
Red-whiskered Bulbul *Pycnonotus jocosus*	紅耳鵯	A	Res	
Chinese Bulbul *Pycnonotus sinensis*	白頭鵯	A	ResMW	
Sooty-headed Bulbul *Pycnonotus aurigaster*	白喉紅臀鵯	C	Res	

Common name	Chinese name	Abundance	Hong Kong Status	World Status
Mountain Bulbul *Ixos mcclellandii*	綠短腳鵯	U	Res	
Chestnut Bulbul *Hemixos castanonotus*	栗背短腳鵯	C	ResW	
Black Bulbul *Hypsipetes leucocephalus*	黑短腳鵯	S	MW	
Brown-eared Bulbul *Hypsipetes amaurotis*	栗耳短腳鵯	V		
Hirundinidae Swallows, Martins				
Grey-throated Martin *Riparia chinensis*	灰喉沙燕	V	-	
Pale Martin *Riparia diluta*	淡色沙燕	U	MW	
Barn Swallow *Hirundo rustica*	家燕	A	MSumW	
Pacific Swallow *Hirundo tahitica*	洋斑燕	V	-	
Common House Martin *Delichon urbicum*	白腹毛腳燕	V	-	
Asian House Martin *Delichon dasypus*	煙腹毛腳燕	U	MW	
Red-rumped Swallow *Cecropis daurica*	金腰燕	C	MWSum	
Pnoepygidae Wren-babblers				
Pygmy Wren-babbler *Pnoepyga pusilla*	小鷯鶥	C	Res	
Cettidae Cettia Bush Warblers and allies				
Rufous-faced Warbler *Abroscopus albogularis*	棕臉鶲鶯	R	W	
Mountain Tailorbird *Phyllergates cuculatus*	金頭縫葉鶯	C	WRes	
Japanese Bush Warbler *Horornis diphone*	日本樹鶯	U	WM	
Manchurian Bush Warbler *Horornis borealis*	遠東樹鶯	S	WM	
Brown-flanked Bush Warbler *Horornis fortipes*	強腳樹鶯	C	WSum	
Asian Stubtail *Urosphena squameiceps*	鱗頭樹鶯	C	W	
Pale-footed Bush Warbler *Urosphena pallidipes*	淡腳樹鶯	V	-	
Aegithalidae Bushtits				
Black-throated Tit *Aegithalos concinnus*	紅頭長尾山雀	S	Res*	
Phylloscopidae Leaf Warblers and allies				
Willow Warbler *Phylloscopus trochilus*	歐柳鶯	V		
Common Chiffchaff *Phylloscopus collybita*	嘰喳柳鶯	V	-	
Dusky Warbler *Phylloscopus fuscatus*	褐柳鶯	A	WM	
Yellow-streaked Warbler *Phylloscopus armandii*	棕眉柳鶯	R	Au	
Radde's Warbler *Phylloscopus schwarzi*	巨嘴柳鶯	U	AuW	
Chinese Leaf Warbler *Phylloscopus yunnanensis*	雲南柳鶯	V	-	
Pallas's Leaf Warbler *Phylloscopus proregulus*	黃腰柳鶯	C	WM	
Yellow-browed Warbler *Phylloscopus inornatus*	黃眉柳鶯	A	WM	
Hume's Leaf Warbler *Phylloscopus humei*	淡眉柳鶯	V	-	
Arctic Warbler *Phylloscopus borealis*	極北柳鶯	C	M	
Japanese Leaf Warbler *Phylloscopus xanthodryas*	日本柳鶯	R	M	
Greenish Warbler *Phylloscopus trochiloides*	暗綠柳鶯	V	-	
Two-barred Warbler *Phylloscopus plumbeitarsus*	雙斑柳鶯	U	MW	
Pale-legged Leaf Warbler *Phylloscopus tenellipes*	淡腳柳鶯	U	MW	
Sakhalin Leaf Warbler *Phylloscopus borealoides*	庫頁島柳鶯	S	M	
Eastern Crowned Warbler *Phylloscopus coronatus*	冕柳鶯	U	MW	
Ijima's Leaf Warbler *Phylloscopus ijimae*	飯島柳鶯	V	-	VU
Goodson's Leaf Warbler *Phylloscopus goodsoni*	古氏[冠紋]柳鶯	C	W	
Emei Leaf Warbler *Phylloscopus emeiensis*	峨眉柳鶯	V	-	
Sulphur-breasted Warbler *Phylloscopus ricketti*	黑眉柳鶯	R	W	-
White-spectacled Warbler *Seicercus affinis*	白眶鶲鶯	R	W	
Grey-crowned Warbler *Seicercus tephrocephalus*	灰冠鶲鶯	V	-	
Bianchi's Warbler *Seicercus valentini*	比氏鶲鶯	V	-	

Common name	Chinese name	Abundance	Hong Kong Status	World Status
Martens's Warbler *Seicercus omeiensis*	峨嵋鶲鶯	V	-	
Alström's Warbler *Seicercus soror*	純色尾鶲鶯	V	-	
Chestnut-crowned Warbler *Seicercus castaniceps*	栗頭鶲鶯	R	W	
Acrocephalidae Reed Warblers and allies				
Oriental Reed Warbler *Acrocephalus orientalis*	東方大葦鶯	C	MW	
Black-browed Reed Warbler *Acrocephalus bistrigiceps*	黑眉葦鶯	C	M	
Blunt-winged Warbler *Acrocephalus concinens*	鈍翅葦鶯	V		
Manchurian Reed Warbler *Acrocephalus tangorum*	遠東葦鶯	S	MW	VU
Paddyfield Warbler *Acrocephalus agricola*	稻田葦鶯	R	WM	
Blyth's Reed Warbler *Acrocephalus dumetorum*	布氏葦鶯	R	WM	
Thick-billed Warbler *Iduna aedon*	厚嘴葦鶯	S	MW	
Booted Warbler *Iduna caligata*	靴籬	V	-	
Sykes's Warbler *Iduna rama*	賽氏篙鶯	V	-	
Locustellidae Grassbirds and allies				
Russet Bush Warbler *Locustella mandelli*	高山短翅鶯	U	WSum	
Baikal Bush Warbler *Locustella davidi*	北短翅鶯	V	-	
Brown Bush Warbler *Locustella luteoventris*	棕褐短翅鶯	R	W	
Lanceolated Warbler *Locustella lanceolata*	矛斑蝗鶯	U	M	
Middendorff's Grasshopper Warbler *Locustella ochotensis*	北蝗鶯	V	-	
Styan's Grasshopper Warbler *Locustella pleskei*	史氏蝗鶯	S	MW	VU
Pallas's Grasshopper Warbler *Locustella certhiola*	小蝗鶯	C	M	
Japanese Swamp Warbler *Locustella pryeri*	斑背大尾鶯	V	-	NT
Cisticolidae Cisticolas and allies				
Zitting Cisticola *Cisticola juncidis*	棕扇尾鶯	C	MWSum	
Golden-headed Cisticola *Cisticola exilis*	金頭扇尾鶯	C	W	
Yellow-bellied Prinia *Prinia flaviventris*	黃腹鷦鶯	A	Res	
Plain Prinia *Prinia inornata*	純色鷦鶯	C	Res	
Common Tailorbird *Orthotomus sutorius*	長尾縫葉鶯	C	Res	
Timaliidae Babblers				
Streak-breasted Scimitar Babbler *Pomatorhinus ruficollis*	棕頸鉤嘴鶥	C	Res*	
Rufous-capped Babbler *Stachyridopsis ruficeps*	紅頭穗鶥	C	Res*	
Pellorneidae Fulvettas, Ground Babblers				
Huet's Fulvetta *Alcippe hueti*	黑眉雀鶥	U	Res*	
Chinese Grassbird *Graminicola striatus*	大草鶯	S	Res	VU
Leiothrichidae Laughingthrushes				
Chinese Babax *Babax lanceolatus*	矛紋草鶥	E	*	
Chinese Hwamei *Garrulax canorus*	畫眉	C	Res	
Masked Laughingthrush *Garrulax perspicillatus*	黑臉噪鶥	A	Res	
Greater Necklaced Laughingthrush *Garrulax pectoralis*	黑領噪鶥	C	Res*	
Black-throated Laughingthrush *Garrulax chinensis*	黑喉噪鶥	C	Res*	
White-browed Laughingthrush *Garrulax sannio*	白頰噪鶥	U	Res*	
Blue-winged Minla *Minla cyanouroptera*	藍翅希鶥	C	Res*	
Silver-eared Leiothrix *Leiothrix argentauris*	銀耳相思鳥	C	Res*	
Red-billed Leiothrix *Leiothrix lutea*	紅嘴相思鳥	U	Res*	
Sylviidae Sylviid Babblers				
Lesser Whitethroat *Sylvia curruca*	白喉林鶯	V	-	
Vinous-throated Parrotbill *Sinosuthora webbiana*	棕頭鴉雀	S	Res*	
Zosteropidae White-eyes				
Chestnut-collared Yuhina *Yuhina torqueola*	栗耳鳳鶥	U	W	

Common name	Chinese name	Abundance	Hong Kong Status	World Status
Chestnut-flanked White-eye *Zosterops erythropleurus*	紅脇繡眼鳥	S	W	
Japanese White-eye *Zosterops japonicus*	暗綠繡眼鳥	A	ResW	
Sittidae Nuthatches				
Velvet-fronted Nuthatch *Sitta frontalis*	絨額鳾	C	Res*	
Sturnidae Starlings				
Crested Myna *Acridotheres cristatellus*	八哥	A	Res	
Common Myna *Acridotheres tristis*	家八哥	C	Res*	
Red-billed Starling *Spodiopsar sericeus*	絲光椋鳥	A	WSum	
White-cheeked Starling *Spodiopsar cineraceus*	灰椋鳥	C	WSum	
Black-collared Starling *Gracupica nigricollis*	黑領椋鳥	C	Res	
Daurian Starling *Agropsar sturninus*	北椋鳥	U	MW	
Chestnut-cheeked Starling *Agropsar philippensis*	栗頰椋鳥	S	M	
White-shouldered Starling *Sturnia sinensis*	灰背椋鳥	C	MSumW	
Chestnut-tailed Starling *Sturnia malabarica*	灰頭椋鳥	R	W	
Rosy Starling *Pastor roseus*	粉紅椋鳥	V	-	
Common Starling *Sturnus vulgaris*	紫翅椋鳥	S	AuW	
Turdidae Thrushes				
Orange-headed Thrush *Geokichla citrina*	橙頭地鶇	S	WMSum	
Siberian Thrush *Geokichla sibirica*	白眉地鶇	S	MW	
White's Thrush *Zoothera aurea*	懷氏地鶇	U	WM	
Grey-backed Thrush *Turdus hortulorum*	灰背鶇	C	WM	
Japanese Thrush *Turdus cardis*	烏灰鶇	C	WM	
Chinese Blackbird *Turdus mandarinus*	中國烏鶇	C	WMRes	
Eyebrowed Thrush *Turdus obscurus*	白眉鶇	U	MW	
Pale Thrush *Turdus pallidus*	白腹鶇	C	W	
Brown-headed Thrush *Turdus chrysolaus*	赤胸鶇	S	WM	
Red-throated Thrush *Turdus ruficollis*	赤頸鶇	V	-	
Naumann's Thrush *Turdus naumanni*	紅尾鶇	R	W	
Dusky Thrush *Turdus eunomus*	斑鶇	S	W	
Chinese Thrush *Turdus mupinensis*	寶興歌鶇	V	-	
Muscicapidae Chats, Old World Warblers				
Oriental Magpie Robin *Copsychus saularis*	鵲鴝	A	Res	
Grey-streaked Flycatcher *Muscicapa griseisticta*	灰紋鶲	U	M	
Dark-sided Flycatcher *Muscicapa sibirica*	烏鶲	U	Au	
Asian Brown Flycatcher *Muscicapa latirostris*	北灰鶲	C	MW	
Brown-breasted Flycatcher *Muscicapa muttui*	褐胸鶲	V	-	
Ferruginous Flycatcher *Muscicapa ferruginea*	棕尾褐鶲	U	M	
Hainan Blue Flycatcher *Cyornis hainanus*	海南藍仙鶲	C	SumMW	
Hill Blue Flycatcher *Cyornis banyumas*	山藍仙鶲	V	-	
Chinese Blue Flycatcher *Cyornis glaucicomans*	中華仙鶲	V	-	
Brown-chested Jungle Flycatcher *Cyornis brunneatus*	白喉林鶲	R	Au	VU
Fujian Niltava *Niltava davidi*	棕腹大仙鶲	S	W	
Small Niltava *Niltava macgrigoriae*	小仙鶲	R	AuW	
Blue-and-white Flycatcher *Cyanoptila cyanomelana*	白腹姬鶲	C	M	
Zappey's Flycatcher *Cyanoptila cumatilis*	琉璃藍鶲	V	-	NT
Verditer Flycatcher *Eumyias thalassinus*	銅藍鶲	U	W	
Lesser Shortwing *Brachypteryx leucophris*	白喉短翅鶇	C	ResW	
Siberian Blue Robin *Larvivora cyane*	藍歌鴝	S	M	
Rufous-tailed Robin *Larvivora sibilans*	紅尾歌鴝	C	WM	

Common name	Chinese name	Abundance	Hong Kong Status	World Status
Japanese Robin *Larvivora akahige*	日本歌鴝	R	W	
Bluethroat *Luscinia svecica*	藍喉歌鴝	C	W	
Siberian Rubythroat *Calliope calliope*	紅喉歌鴝	C	WM	
White-tailed Robin *Myiomela leucura*	白尾藍地鴝	V	-	
Red-flanked Bluetail *Tarsiger cyanurus*	紅脇藍尾鴝	C	WM	
Slaty-backed Forktail *Enicurus schistaceus*	灰背燕尾	R	WSum	
Blue Whistling Thrush *Myophonus caeruleus*	紫嘯鶇	C	Res	
Yellow-rumped Flycatcher *Ficedula zanthopygia*	白眉姬鶲	U	Au	
Narcissus Flycatcher *Ficedula narcissina*	黃眉姬鶲	U	M	
Green-backed Flycatcher *Ficedula elisae*	綠背姬鶲	V	-	
Mugimaki Flycatcher *Ficedula mugimaki*	鴝姬鶲	U	MW	
Slaty-backed Flycatcher *Ficedula hodgsonii*	銹胸藍姬鶲	V	-	
Rufous-gorgeted Flycatcher *Ficedula strophiata*	橙胸姬鶲	R	W	
Red-breasted Flycatcher *Ficedula parva*	紅胸姬鶲	S	MW	
Taiga (Red-throated) Flycatcher *Ficedula albicilla*	紅喉姬鶲	C	MW	
Black Redstart *Phoenicurus ochruros*	赭紅尾鴝	V	-	
Hodgson's Redstart *Phoenicurus hodgsoni*	黑喉紅尾鴝	V	-	
Daurian Redstart *Phoenicurus auroreus*	北紅尾鴝	C	W	
Plumbeous Water Redstart *Phoenicurus fuliginosus*	紅尾水鴝	U	W	
Blue Rock Thrush *Monticola solitarius*	藍磯鶇	C	MWSum	
Chestnut-bellied Rock Thrush *Monticola rufiventris*	栗腹磯鶇	R	W	
White-throated Rock Thrush *Monticola gularis*	白喉磯鶇	R	MW	
Stejneger's Stonechat *Saxicola stejnegeri*	黑喉石(即鳥)	C	MW	
Grey Bush Chat *Saxicola ferreus*	灰林(即鳥)	S	WM	
Pied Wheatear *Oenanthe pleschanka*	白頂(即鳥)	V	-	
Chloropseidae Leafbirds				
Orange-bellied Leafbird *Chloropsis hardwickii*	橙腹葉鵯	U	ResW	
Dicaeidae Flowerpeckers				
Plain Flowerpecker *Dicaeum minullum*	純色啄花鳥	S	W	
Fire-breasted Flowerpecker *Dicaeum ignipectus*	紅胸啄花鳥	U	W	
Scarlet-backed Flowerpecker *Dicaeum cruentatum*	朱背啄花鳥	C	Res	
Nectariniidae Sunbirds				
Olive-backed Sunbird *Cinnyris jugularis*	黃腹花蜜鳥	V		
Mrs Gould's Sunbird *Aethopyga gouldiae*	藍喉太陽鳥	R	W	
Fork-tailed Sunbird *Aethopyga christinae*	叉尾太陽鳥	C	ResW	
Passeridae Old World Sparrows				
House Sparrow *Passer domesticus*	家麻雀	V	-	
Russet Sparrow *Passer rutilans*	山麻雀	V	-	
Eurasian Tree Sparrow *Passer montanus*	樹麻雀	A	Res	
Ploceidae Weavers, Widowbirds				
Baya Weaver *Ploceus philippinus*	黃胸織雀	E	*	
Estrildidae Waxbills, Munias and allies				
White-rumped Munia *Lonchura striata*	白腰文鳥	C	Res	
Scaly-breasted Munia *Lonchura punctulata*	斑文鳥	A	Res	
Chestnut Munia *Lonchura atricapilla*	栗腹文鳥	E	*	
Motacillidae Wagtails, Pipits				
Forest Wagtail *Dendronanthus indicus*	山鶺鴒	U	MW	
Eastern Yellow Wagtail *Motacilla tschutschensis*	東黃鶺鴒	C	MW	

Common name	Chinese name	Abundance	Hong Kong Status	World Status
Citrine Wagtail *Motacilla citreola*	黃頭鶺鴒	U	WM	
Grey Wagtail *Motacilla cinerea*	灰鶺鴒	C	WM	
White Wagtail *Motacilla alba*	白鶺鴒	C	MWSum	
Richard's Pipit *Anthus richardi*	理氏鷚	C	MWRes	
Blyth's Pipit *Anthus godlewskii*	布氏鷚	V	-	
Olive-backed Pipit *Anthus hodgsoni*	樹鷚	C	WM	
Pechora Pipit *Anthus gustavi*	北鷚	S	M	
Rosy Pipit *Anthus roseatus*	粉紅胸鷚	V	-	
Red-throated Pipit *Anthus cervinus*	紅喉鷚	C	MW	
Buff-bellied Pipit *Anthus rubescens*	黃腹鷚	U	MW	
Water Pipit *Anthus spinoletta*	水鷚	V	-	
Upland Pipit *Anthus sylvanus*	山鷚	U	Res	
Fringillidae Finches				
Brambling *Fringilla montifringilla*	燕雀	S	MW	
Hawfinch *Coccothraustes coccothraustes*	錫嘴雀	R	W	
Chinese Grosbeak *Eophona migratoria*	黑尾蠟嘴雀	C	WSum	
Japanese Grosbeak *Eophona personata*	黑頭蠟嘴雀	R	W	
Common Rosefinch *Carpodacus erythrinus*	普通朱雀	S	WM	
Grey-capped Greenfinch *Chloris sinica*	金翅雀	S	Res	
Eurasian Siskin *Spinus spinus*	黃雀	S	W	
Emberizidae Buntings				
Crested Bunting *Emberiza lathami*	鳳頭鵐	R	M	
Slaty Bunting *Emberiza siemsseni*	藍鵐	V	-	
Pine Bunting *Emberiza leucocephalus*	白尾藍地鴝	V	-	
Ortolan Bunting *Emberiza hortulana*	圃鵐	V	-	
Tristram's Bunting *Emberiza tristrami*	白眉鵐	U	W	
Chestnut-eared Bunting *Emberiza fucata*	栗耳鵐	U	MW	
Little Bunting *Emberiza pusilla*	小鵐	C	WM	
Yellow-browed Bunting *Emberiza chrysophrys*	黃眉鵐	S	M	
Rustic Bunting *Emberiza rustica*	田鵐	R	W	
Yellow-throated Bunting *Emberiza elegans*	黃喉鵐	R	M	
Yellow-breasted Bunting *Emberiza aureola*	黃胸鵐	C	AuW	EN
Chestnut Bunting *Emberiza rutila*	栗鵐	U	MW	
Black-headed Bunting *Emberiza melanocephala*	黑頭鵐	S	AuW	
Red-headed Bunting *Emberiza bruniceps*	褐頭鵐	V	-	
Japanese Yellow Bunting *Emberiza sulphurata*	硫磺鵐	S	Sp	VU
Black-faced Bunting *Emberiza spodocephala*	灰頭鵐	C	MW	
Pallas's Reed Warbler *Emberiza pallasi*	葦鵐	R	Au	
Japanese Reed Bunting *Emberiza yessoensis*	紅頸葦鵐	V	-	NT
Common Reed Bunting *Emberiza schoeniclus*	蘆鵐	R	W	

■ Further Information ■

Societies

Hong Kong Bird Watching Society
7C V Ga Building
533 Castle Peak Road
Lai Chi Kok, Kowloon
www.hkbws.org.hk

The Hong Kong Natural History Society
www.hknhs.org

Kadoorie Farm and Botanic Gardens
Lam Kam Road
Tai Po, New Territories
www.kfbg.hk

World Wide Fund for Nature
1 Tramway Path
Central
Hong Kong
and
15/F Manhattan Centre
8 Kwai Chung Road
Kwai Chung, New Territories

Useful websites

www.hkbws.org.hk
The website of The Hong Kong Bird Watching Society provides extensive information on Hong Kong birdwatching sites, public transport, the Hong Kong Bird List, research reports, meetings and outings.

www.hko.gov.hk/tide/estation_select.htm
The website of The Hong Kong Observatory offers tide predictions at selected Hong Kong locations.

www.hko.gov.hk/wxinfo/currwx/fnd.htm
The website of The Hong Kong Observatory gives daily weather bulletins and forecasts, and a nine-day forecast for Hong Kong.

www.orientalbirdimages.org
An extensive library of images of birds of the Oriental region including Hong Kong species.

www.xeno-canto.org
An extensive library of sound recordings of bird calls and songs including Hong Kong species.

■ Further Information ■

Books

Brazil, M. 2009. *Birds of East Asia*. Christopher Helm. London.
Identification guide to the birds of Eastern China (including Hong Kong), Taiwan, Korea, Japan and Eastern Russia.

Carey, G.J. (Ed). 2001. *The Avifauna of Hong Kong*. Hong Kong Bird Watching Society. Hong Kong.
Comprehensive and authoritative,
detailed, annotated checklist of the birds
of Hong Kong.

Diskin, D. 2014. *Mai Po: the Seasons*. Accipiter Press. Hong Kong.
Illustrates a year at Mai Po in pictures of its physical features, and flora and fauna, particularly birds.

The Hong Kong Bird Watching Society. *Hong Kong Bird Reports*.
Annual reports since 1958.

The Hong Kong Bird Watching Society. 2012. *A Photographic Guide to the Birds of Hong Kong*. Wan Li Book Co. Ltd. Hong Kong.
Photographic guide illustrating 456 species. Bilingual text.

Viney, C.A., Phillipps, K. & Lam, C.Y. 2005. *The Birds of Hong Kong and South China*. Eighth edition.
International Services Department.
Hong Kong.
The standard field guide to the birds of Hong Kong; also covers South China.

INDEX

Accipiter gularis 35
  soloensis 35
  trivirgatus 34
  virgatus 36
Acridotheres cristatellus 127
  tristis 128
Acrocephalus bistrigiceps 116
  orientalis 115
Actitis hypoleucos 56
Aethopyga christinae 146
Agropsar sturninus 130
Alauda arvensis 103
Alcedo atthis 84
Alcippe hueti 121
Amaurornis phoenicurus 40
Anas acuta 18
  clypeata 18
  crecca 19
  falcata 19
  penelope 17
  querquedula 19
  strepera 16
  zonorhyncha 17
Anthus cervinus 153
  hodgsoni 152
  richardi 152
  rubescens 153
  sylvanus 154
Apus nipalensis 82
  pacificus 82
Aquila fasciata 34
  heliaca 33
Ardea alba 28
  cinerea 27
  purpurea 28
Ardeola bacchus 26
Arenaria interpres 57
Avocet, Pied 42
Aythya fuligula 20
Babbler, Rufous-capped 121
  Streak-breasted Scimitar 120
Barbet, Great 86
Bee-eater, Blue-tailed 85
Besra 36
Bittern, Cinnamon 25
  Eurasian 24
  Yellow 24
Blackbird, Chinese 132
Bluetail, Red-flanked 140
Bluethroat 139
Boobook, Northern 81
Botaurus stellaris 24
Brachypteryx leucophris 138
Bubulcus coromandus 27
Bulbul, Chestnut 106
  Chinese 104
  Mountain 105
  Red-whiskered 104
  Sooty-headed 105
Bunting, Black-faced 157
  Chestnut 157
  Chestnut-eared 155
  Little 156

  Tristram's 155
  Yellow-breasted 156
Butastur indicus 38
Buteo japonicus 88
Butorides striata 26
Buzzard, Crested Honey 32
  Eastern 39
  Grey-faced 38
Cacatua sulphurea 89
Cacomantis merulinus 78
Calidris acuminata 61
  alba 58
  alpina 62
  canutus 58
  ferruginea 61
  minuta 59
  ruficollis 59
  subminuta 60
  temminckii 60
  tenuirostris 57
Calliope calliope 139
Canary-flycatcher, Grey-headed 101
Caprimulgus affinis 81
Cecropis daurica 108
Centropus bengalensis 76
  sinensis 76
Ceryle rudis 85
Chalcophaps indica 75
Charadrius alexandrinus 45
  dubius 44
  leschenaultii 46
  mongolus 45
Chlidonias hybrida 71
  leucopterus 71
Chloropsis hardwickii 145
Chroicocephalus ridibundus 64
  saundersi 65
Circus melanoleucos 37
  spilonotus 36
Cisticola exilis 118
  juncidis 118
Cisticola, Golden-headed 118
  Zitting 118
Clamator coromandus 77
Clanga clanga 33
Cockatoo, Yellow-crested 89
Columba livia 73
Coot, Eurasian 41
Copsychus saularis 134
Coracina melaschistos 91
Cormorant, Great 30
Coturnix japonica 21
Corvus macrorynchos 101
  splendens 100
  torquatus 100
Coucal, Greater 76
  Lesser 76
Crake, Ruddy-breasted 40
  Slaty-legged 39
Crow, Collared 100
  House 100
  Large-billed 101

Cuckoo, Chestnut-winged 77
  Hodgson's Hawk 79
  Indian 79
  Large Hawk 78
  Plaintive 78
Cuckooshrike, Black-winged 91
Cuculus micropterus 79
Culicicapa ceylonensis 101
Curlew, Eurasian 51
  Far Eastern 51
Cyanopica cyanus 98
Cyanoptila cyanomelana 137
Cyornis hainanus 136
Delichon dasypus 107
Dendrocitta formosae 99
Dendronanthus indicus 148
Dicaeum cruentatum 146
  ignipectus 145
Dicrurus hottentottus 96
  leucophaeus 95
  macrocercus 95
Dollarbird, Oriental 83
Dove, Common Emerald 75
  Eurasian Collared 74
  Oriental Turtle 73
  Red Turtle 74
  Spotted 75
Dowitcher, Asian 49
Drongo, Ashy 95
  Black 95
  Hair-crested 96
Duck, Chinese Spot-billed 17
  Falcated 16
  Tufted 20
Dunlin 62
Eagle, Bonelli's 34
  Crested Serpent 32
  Eastern Imperial 33
  Greater Spotted 33
  White-bellied Sea 38
Egret, Eastern Cattle 27
  Great 28
  Intermediate 29
  Little 29
Egretta garzetta 29
  intermedia 29
  sacra 30
Elanus caeruleus 31
Emberiza aureola 156
  fucata 155
  pusilla 156
  rutila 157
  spodocephala 157
  tristrami 155
Eophona migratoria 154
Erpornis, White-bellied 94
Erpornis zantholeuca 94
Eudynamys scolopaceus 77
Eumyias thalassinus 137
Eurystomus orientalis 83
Falco amurensis 88
  peregrinus 89
  subbuteo 88

INDEX

tinnunculus 87
Falcon, Amur 88
  Peregrine 89
Ficedula albicilla 142
  mugimaki 142
  narcissina 141
  zanthopygia 141
Flowerpecker, Fire-breasted 145
  Scarlet-backed 146
Flycatcher, Asian Brown 135
  Blue-and-white 137
  Dark-sided 135
  Ferruginous 136
  Grey-streaked 134
  Hainan Blue 136
  Mugimaki 142
  Narcissus 141
  Taiga (Red-throated) 142
  Verditer 137
  Yellow-rumped 141
Francolin, Chinese 20
Francolinus pintadeanus 20
Fulvetta, Huet's 121
Fulica atra 41
Gadwall 16
Gallinago gallinago 48
  stenura 48
Gallinula chloropus 41
Garganey 19
Garrulax canorus 122
  chinensis 123
  pectoralis 123
  perspicillatus 122
  sannio 124
Gelochelidon nilotica 66
Glareola maldivarum 63
Glaucidium cuculoides 80
Godwit, Bar-tailed 50
  Black-tailed 49
Goshawk, Crested 34
Gracupica nigricollis 129
Grebe, Great Crested 22
  Little 22
Greenshank, Common 53
  Nordmann's 54
Grosbeak, Chinese 154
Gull, Black-headed 64
  Black-tailed 65
  Caspian 65
  Heuglin's 66
  Saunders's 64
Halcyon pileata 84
  smyrnensis 83
Haliaeetus leucogaster 38
Harrier, Eastern Marsh 36
  Pied 37
Hemixos castanonotus 106
Heron, Black-crowned Night 25
  Chinese Pond 26
  Grey 27
  Pacific Reef 30
  Purple 28
  Striated 26

Hierococcyx nisicolor 79
  sparverioides 78
Himantopus himantopus 42
Hirundo rustica 107
Hobby, Eurasian 88
Hoopoe, Eurasian 86
Horornis diphone 109
  fortipes 110
Hwamei, Chinese 122
Hydrophasianus chirurgus 47
Hydroprogne caspia 67
Hypothymis azurea 96
Ixobrychus cinnamomeus 25
  sinensis 24
Ixos mcclellandii 105
Jacana, Pheasant-tailed 47
Jaeger, Long-tailed 72
Jynx torquilla 87
Kestrel, Common 87
Kingfisher, Black-capped 84
  Common 84
  Pied 85
  White-throated 83
Kite, Black 37
  Black-winged 31
Knot, Great 57
  Red 56
Koel, Asian 77
Lanius cristatus 93
  schach 93
Lapwing, Grey-headed 43
Larus cachinnans 65
  crassirostris 65
  fuscus heuglini 66
Larvivora sibilans 138
Laughingthrush, Black-throated 123
  Greater Necklaced 123
  Masked 122
  White-browed 124
Leafbird, Orange-bellied 145
Leiothrix argentauris 125
  lutea 125
Leiothrix, Red Billed 125
Limicola falcinellus 62
Limnodromus semipalmatus 49
Limosa lapponica 50
  limosa 49
Locustella certhiola 117
  lanceolata 117
  mandelli 116
Lonchura punctulata 148
  striata 147
Luscinia svecica 139
Machlolophus spilonotus 102
Magpie, Azure-winged 98
  Eurasian 99
  Red-billed Blue 98
Martin, Asian House 107
  Pale 106
Merops philippinus 85
Mesia, Silver-eared 125
Milvus migrans 37

Minivet, Ashy 91
  Grey-chinned 92
  Scarlet 92
Minla, Blue-winged 124
Minla cyanouroptera 124
Monarch, Black-naped 96
Monticola solitarius 144
Moorhen, Common 41
Motacilla alba 151
  cinerea 151
  citreola 150
  tschutschensis 149
Munia, Scaly-breasted 148
  White-rumped 147
Murrelet, Ancient 72
Muscicapa ferruginea 136
  griseisticta 134
  latirostris 135
  sibirica 135
Myna, Common 128
  Crested 127
Myophonus caeruleus 140
Nightjar, Savanna 81
Ninox japonica 81
Numenius arquata 51
  madagascariensis 51
  phaeopus 50
Nuthatch, Velvet-fronted 147
Nycticorax nycticorax 25
Onychoprion aleuticus 68
  anaethetus 69
Oriole, Black-naped 94
  Oriolus chinensis 94
Orthotomus sutorius 120
Otis lettia 80
Osprey, Western 31
Owlet, Asian Barred 80
Owl, Collared Scops 80
Painted-snipe, Greater 46
Pandion haliaetus 31
Paradise-Flycatcher, Amur 97
  Japanese 97
Parakeet, Alexandrine 90
  Rose-ringed 90
Parus cinereus 102
Passer montanus 101
Pericrocotus divaricatus 91
  solaris 92
  speciosus 92
Pernis ptilorhyncus 32
Phalacrocorax carbo 30
Phalarope, Red-necked 63
Phalaropus lobatus 63
Phoenicurus auroreus 143
  fuliginosus 143
Phyllergates cucullatus 109
Phylloscopus borealis 113
  coronatus 114
  fuscatus 111
  goodsoni 115
  inornatus 112
  plumbeitarsus 113
  proregulus 112

INDEX

schwarzi 111
tenellipes 114
Pica pica 99
Pigeon, Domestic 73
Pintail, Northern 18
Pipit, Buff-bellied 153
　Olive-backed 152
　Red-throated 153
　Richard's 152
　Upland 154
Platalea leucorodia 23
　minor 23
Plover, Greater Sand 46
　Grey 44
　Kentish 45
　Lesser Sand 45
　Little Ringed 44
　Pacific Golden 43
Pluvialis fulva 43
　squatarola 44
Pnoepyga pusilla 108
Podiceps cristatus 22
Pomatorhinus ruficollis 120
Porzana fusca 40
Pratincole, Oriental 63
Prinia flaviventris 119
　inornata 119
Prinia, Plain 119
　Yellow-bellied 119
Psilopogon virens 86
Psittacula eupatria 90
　krameri 90
Puffinus tenuirostris 21
Pycnonotus aurigaster 105
　jocosus 104
　sinensis 104
Quail, Japanese 21
Rallina eurizonoides 39
Recurvirostra avosetta 42
Redshank, Common 52
　Spotted 52
Redstart, Daurian 143
　Plumbeous Water 143
Remiz consobrinus 103
Riparia diluta 106
Robin, Oriental Magpie 134
　Rufous-tailed 138
Rostratula benghalensis 46
Rubythroat, Siberian 139
Sanderling 58
Sandpiper, Broad-billed 62
　Common 56
　Curlew 61
　Green 54
　Marsh 53
　Sharp-tailed 61
　Terek 56
　Wood 55
Saxicola stejnegeri 144
Scolopax rusticola 47
Shearwater, Short-tailed 21
Shortwing, Lesser 138
Shoveler, Northern 18

Shrike, Brown 93
　Long-tailed 93
Sitta frontalis 127
Skylark, Eurasian 103
Snipe, Common 48
　Pintail 48
Sparrow, Eurasian Tree 147
Sparrowhawk, Chinese 35
　Japanese 35
Spilopelia chinensis 75
Spilornis cheela 32
Spodiopsar cineraceus 129
　sericeus 128
Spoonbill, Black-faced 23
　Eurasian 23
Stachyridopsis ruficeps 121
Starling, Black-collared 129
　Daurian 130
　Red-billed 128
　White-cheeked 129
　White-shouldered 130
Stercorarius longicaudus 72
Sterna dougallii 69
　hirundo 70
　sumatrana 70
Sternula albifrons 68
Stilt, Black-winged 42
Stint, Little 59
　Long-toed 60
　Red-necked 59
　Temminck's 60
Stonechat, Stejneger's 144
Streptopelia decaocto 74
　orientalis 73
　tranquebarica 74
Stubtail, Asian 110
Sturnia sinensis 130
Sunbird, Fork-tailed 146
Swallow, Barn 107
　Red-rumped 108
Swift, House 82
　Pacific 82
Synthliboramphus antiquus 72
Tachybaptus ruficollis 22
Tailorbird, Common 120
　Mountain 109
Tarsiger cyanurus 140
Tattler, Grey-tailed 55
Teal, Eurasian 17
Tern, Aleutian 68
　Black-naped 70
　Bridled 69
　Caspian 67
　Common 70
　Greater Crested 67
　Gull-billed 66
　Little 68
　Roseate 69
　Whiskered 71
　White-winged 71
Terpsiphone atrocaudata 97
　incei 97
Thalasseus bergii 67

Thrush, Blue Rock 144
　Blue Whistling 140
　Eyebrowed 133
　Grey-backed 131
　Japanese 132
　Pale 133
　White's 131
Tit, Chinese Penduline 103
　Cinereous 102
　Yellow-cheeked 102
Treepie, Grey 99
Tringa brevipes 55
　erythropus 52
　glareola 55
　guttifer 54
　nebularia 53
　ochropus 54
　stagnatilis 53
　totanus 52
Turdus cardis 132
　hortulorum 131
　mandarinus 132
　obscurus 133
　pallidus 133
Turnstone, Ruddy 57
Upupa epops 86
Urocissa erythrorhyncha 98
Urosphena squameiceps 110
Vanellus cinereus 43
Wagtail, Eastern Yellow 149
　Citrine 150
　Forest 148
　Grey 151
　White 151
Warbler, Arctic 113
　Black-browed Reed 116
　Brown-flanked Bush 110
　Dusky 111
　Eastern Crowned 114
　Goodson's Leaf 115
　Japanese Bush 109
　Lanceolated 117
　Oriental Reed 115
　Pale-legged Leaf 114
　Pallas's Grasshopper 117
　Pallas's Leaf 112
　Radde's 111
　Russet Bush 116
　Two-barred 113
　Yellow-browed 112
Waterhen, White-breasted 40
Whimbrel 50
Wigeon, Eurasian 17
Woodcock, Eurasian 47
Wren-babbler Pygmy 108
Wryneck, Eurasian 87
White-eye, Japanese 126
Xenus cinereus 56
Yuhina, Chestnut-collared 126
Yuhina torqueola 126
Zoothera aurea 131
Zosterops japonicus 126